TETHERED FATES

TETHERED FATES

Companies, Communities, and
Rights at Stake

SHAREEN HERTEL

Oxford University Press is a department of the University of Oxford. It furthers
the University's objective of excellence in research, scholarship, and education
by publishing worldwide. Oxford is a registered trade mark of Oxford University
Press in the UK and certain other countries.

Published in the United States of America by Oxford University Press
198 Madison Avenue, New York, NY 10016, United States of America.

© Oxford University Press 2019

All rights reserved. No part of this publication may be reproduced, stored in
a retrieval system, or transmitted, in any form or by any means, without the
prior permission in writing of Oxford University Press, or as expressly permitted
by law, by license, or under terms agreed with the appropriate reproduction
rights organization. Inquiries concerning reproduction outside the scope of the
above should be sent to the Rights Department, Oxford University Press, at the
address above.

You must not circulate this work in any other form
and you must impose this same condition on any acquirer.

CIP data is on file at the Library of Congress
ISBN 978-0-19-090384-8 (pbk.)
ISBN 978-0-19-090383-1 (hbk.)

9 8 7 6 5 4 3 2 1

Paperback printed by Webcom, Inc., Canada
Hardback printed by Bridgeport National Bindery, Inc., United States of America

*For my father, Phil Hertel,
who never stops teaching*

CONTENTS

	Acknowledgments	ix
1.	Introduction	1
2.	A Genealogy of Community Consultation	15
3.	Global Landscape and Local Contexts for Stakeholder Consultation	36
4.	The People Beyond the Tag: Stakeholder Perceptions in Villa Altagracia	61
5.	Challenges Down the Road: Stakeholder Perceptions in Bonao	98
6.	Policy Implications of Changes in Stakeholder Consultation	133
	Epilogue	163
	Appendix 1: Interview Questionnaire	*167*
	Appendix 2: Cuestionario	*171*
	Notes	*175*
	References	*191*
	Index	*211*

ACKNOWLEDGMENTS

This book would not have been possible without the support of the many institutions and people named here, along with others who have shared their stories or supported me in practical ways on a daily basis as I conducted research and traveled but whom I cannot name. In the process, they have collectively helped me bring this project to fruition.

At the University of Connecticut, I am grateful for research grants from the Human Rights Institute (HRI), the Office of the Provost (Academic Plan Award for Business and Human Rights Engaged Research), and the Office of the Vice President for Research (Scholarship Facilitation Fund). The Department of Political Science, The Thomas J. Dodd Center, and HRI have all offered superb administrative support; special thanks to Lindsay Halle, Nana Amos, and Lyndsay Nalbandian.

My academic colleagues have shared insights on various portions of the research, and I am grateful to those who generously commented in workshops and colloquia at the University of Connecticut (through the HRI Economic and Social Rights Program and the Institute for Latina/o, Caribbean and Latin American Studies) as well as colleagues at New York University School of Law Center for Human Rights and Global Justice; the Orfalea Center for Global and International Studies at the University of California Santa Barbara; Columbia University's "University Seminar" series; the Department of Political Science at Binghamton University; and in panels over several years at the annual meetings of the International Studies Association and the American Political Science Association. I am deeply grateful to Dawn Brancati and Kathryn Libal for their willingness to read and comment on the manuscript.

I have also been fortunate to receive comments on this research from colleagues in the policy, advocacy, and business communities who have been involved in program activity under our UConn Business and Human Rights Initiative. Kyle Muncy and Liz Kennedy have offered an on-the-ground perspective from within the collegiate licensing world. I also continue to draw inspiration from the work of the many advocates who have been involved in this initiative from around the world, and whose work pushes the frontiers of social justice in global supply chains. I admire your candor and courage.

On a practical level, the research in this book would not have been possible without the support of gifted research assistants Lauren Pérez-Bonilla, Ariana Javidi, Emily Friedman, and Rajeshwari Majumdar. Elizabeth Mahan has provided superb editorial support. Colleagues at the Business and Human Rights Resource Centre offered timely and expert advice on the use of their data and have adapted it for use by me and other scholars. Allison Petrozziello, in Santo Domingo, provided invaluable intellectual and practical guidance for my field research in the Dominican Republic. Community members and their allies in the towns of Villa Altagracia and Bonao (in the neighborhood of Las Delicias) made the case study chapters possible through their generosity and courage in sharing their truth. Angela Chnapko at Oxford University Press moved the book forward from start to finish with grace.

Since human rights begin at home, as Eleanor Roosevelt said, I also owe a huge debt to my husband, Donald Swinton. His critical feedback shaped the manuscript, as did our shared belief that a project straddling constituencies, paradigms, and places was not only possible but needed. Finally, I thank our daughter, Jo, for her patience with my divided attention, for her willingness to burden-share when I am away, and for her extraordinary will and determination to make her own contribution to the world in her own way. It is her stake in the future that makes this book on stakeholder dialogue matter the most.

TETHERED FATES

CHAPTER 1

Introduction

OVERVIEW

The principle that people in communities affected by a company's production activities have the right to remedy if they are harmed by a company's activity—even if they do not personally work on a farm or a factory where goods are produced—is an intuitively sensible idea but a challenging one to translate into practice. As various waves of contemporary global economic integration have unfolded beginning in the 1970s and then with renewed vigor in the 1990s, a host of actors from labor unions, consumer and advocacy organizations, and international bodies such as the International Labour Organization have shed light on the abuse of workers who make the products we use and the rights of people in communities affected by activity in global supply chains (Ramasastry 2015; Bauer 2011, 182–185; Jerbi 2009).

Businesses have developed more sophisticated systems for preventing "sweatshop" labor in their supply chains, along with improvements in the environmental sustainability of product design, production, and end-use. Parallel to changes in the global marketplace have come changes in human rights theory and practice. These include the emergence of a "business and human rights" (BHR) set of rules, standards, and institutions formalized through the United Nations system in the early 2000s and increasingly taken up by international development banks, governments, leading companies, and civil society organizations (Bird, Cahoy, and Prenkert 2014). The "stakeholder dialogue" central to contemporary BHR is grounded in the understanding that businesses, governments, and civil society are supposed to be *jointly* responsible for shaping the remedies available to people harmed (directly or indirectly) in the course of business activity, wherever it takes place.[1]

Yet as this book will demonstrate, the theoretical underpinnings of remedy remain underdeveloped,[2] and multi-stakeholder initiatives (MSIs) and other forms of dialogue initiated by companies as a vehicle for determining remedy are often poorly executed in practice. In many cases, community members are left out of the process of identifying not only risk but also potential ways that companies can add social value in the communities where they operate globally. Grassroots-level frustration with the limitations of BHR has led to the emergence of an alternative "worker-driven social responsibility" (WSR) paradigm developed by workers and community-based allies who challenge both the legitimacy and effectiveness of conventional BHR. They are forging binding standards and mechanisms for monitoring and enforcing corporate responsibility from the bottom up.

This book comes at a critical juncture, when the stakes for people involved in and affected by global business are high, and popular frustration with persistent shortfalls in economic rights fulfillment looms large. The core challenge of finding structural remedies for economic rights violations is bigger than just identifying the individual disputes between a particular factory and a particular group of workers or community members. As the chapters that follow demonstrate, members of local communities often do not perceive themselves as having grounds for remedy yet are impacted (directly or indirectly) by company actions. The book thus argues that the responsibility for remedy needs to be built into the business model as a cost of doing business in global supply chains. One of the book's main contributions is to harness the notion of remedy to address the structural challenge of underdevelopment in factory communities. Another contribution is to harness multiple methods and multiple sources of data (e.g., historical, statistical, interview, and participant observation–based data) to uncover the challenges of stakeholder consultation in theory and practice.

As this book demonstrates, the contemporary stakeholder consultation taking place between companies and stakeholders globally typically happens in extractive industries (e.g., oil, gas, and mining) where the sunk costs of operations are high, not in places where light manufacturing takes place and the threat of exit looms large (Hirschman 1970). When they do occur, consultations often take place at a time and place more convenient for companies than for community members. Nor are the types of remedies proposed necessarily well integrated with existing government programs. Nor are those remedies offered always helpful to community members.

Empirically, the book explores the conditions under which stakeholder dialogue occurs and the types of remedies available to communities in one of the most challenging yet comparatively under researched sectors: namely, in communities where light manufacturing dominates the economy. Jobs in light manufacturing, particularly in the textile industry (the world's most widely dispersed industrial sector), are often the first type of formal employment for many poor people (Bair, Dickson, and Miller 2014, 3). Even people not employed in a factory often hope for the broader economic benefits that light manufacturing activity can bring to their communities.

Yet weak governance on the part of states exacerbates downward pressure on wages and working conditions in many highly mobile industries. Even people not directly employed in light manufacturing risk both present externalities (e.g., pollution from manufacturing) and future ones if companies relocate from their communities to places where labor is cheaper or regulations are even less stringent (which can depress the local economy more generally). Indeed, the speed with which such companies can shift production contexts (Rivoli 2015) complicates the prospects for stakeholder dialogue in light manufacturing, as does uneven access to publicly available information on corporate practices at the local level. Poverty as well as complex social pressures faced by people in local communities where factories are sited (Nadvi 2004; Workers Rights Consortium 2013) can also diminish the ability or willingness of people at the grassroots level to take part in stakeholder consultation.

There is little research that analyzes the emergence and implications of stakeholder consultation in terms of how this practice contributes to (or detracts from) fulfilling the economic rights of people who work and live in communities around the world where light manufacturing takes place. Economic rights include the right to subsistence, the right to work, and the right to income guarantees for those who cannot work (Hertel and Minkler 2007). States, private-sector actors, and people ourselves individually and collectively (through labor unions) bear varying types of responsibility for fulfilling different aspects of economic rights. The state has an obligation not to violate rights directly itself, and to protect against harm by regulating the activities of non-state actors. Private-sector actors have the responsibility to respect local and national laws, at minimum, in the settings where they do business,[3] and to provide remedy when harm occurs. Individual people have the right of access to economic rights without discrimination as well as the duty to contribute to fulfilling our own economic rights through our efforts.

The protection of economic rights and their greater fulfillment over time thus hinges on the ability of state, corporate, and civil society–based actors to broker compromise over burden-sharing and resource allocation. The right to remedy is necessary because this balancing act is not always successful, and the weakest are often harmed the most when it fails. Stakeholder consultation, in principle, aims at enabling people harmed by corporate action to have a say in setting the scope of remedy. But there is little research that unpacks the process from the perspective of community members themselves.[4] This book fills these critical gaps by asking: *Can stakeholder dialogue lead to structural remedies for the economic rights violations faced by people who live in communities where light manufacturing is prevalent? Why or why not, and under what conditions?*

CONCEPTUAL FRAMEWORK

Stakeholder dialogue is a highly scripted process of naming wrongs and claiming rights that draws civil society and business actors directly into contact. The term implies a horizontal relationship among people affected by business activity because it extends the scope of claim-making on companies beyond the traditional stockholder (i.e., a person with legal claim to influence corporate action by virtue of holding stock in it) to include a broader range of people affected by the company's action. But hierarchy is intrinsic to the process. At the most immediate level, the initiator of the formal process of dialogue typically remains the company. With that role as convener (and ultimate arbiter of settlements) comes the power to set the scope of the conversation—both its temporal framework and the nature of the impact that words will have on action.

At a broader level, stakeholder dialogue is tied to struggles over the social legitimacy of BHR. States set a normative framework for BHR by signing and ratifying international treaty law, or by endorsing soft law and nonbinding standards such as the UN Guiding Principles on Business and Human Rights (UN Human Rights Council 2011). States also have the power to create and enforce national laws/regulatory policies that can both constrain business activity and set thresholds for claim-making by people in civil society. But violations of human rights persist in large part because of a lack of state capacity or political will to protect and promote rights. Corporations have become increasingly attuned to the risk that weak state governance and corresponding rights violations pose to business reputation in large part because much light manufacturing takes place in settings where governance is weak. Companies have thus developed a host of voluntary compliance measures ranging from individual company codes of conduct to industry-wide monitoring standards, to product labeling programs grounded in specific labor or environmental standards (Büthe and Mattli 2011) and have sought social buy-in through stakeholder engagement in order to shore up the social legitimacy of these alternative governance processes.

But the assumption of legitimacy intrinsic to stakeholder dialogue is problematic on several levels. Some scholars interpret corporate involvement in BHR (within which stakeholder dialogue occurs) as the latest effort by companies to constrain the scope of emerging norms, the terms of formal government regulation, and the depth of social engagement in civil regulation (Kaplan 2015; Kinderman 2012). For example, Mwangi, Rieth, and Schmitz (2013) along with Deitelhoff and Wolf (2013) argue that when companies take the lead in promoting rights, they are often motivated by competitive interests. When denial of human rights responsibilities no longer works—either because of consumer or industry pressure, or because of the lack of state capacity or willingness to safeguard rights in particularly difficult production settings—companies become human rights norms protagonists in order to "remain ahead of stakeholder and public expectations or to reduce competitive losses that could result from the compliance with human rights standards of some but not all competitors in a given market segment" (2013, 237, and 232–234; see also Zadek 2004).

Within the constructivist international relations literature on the evolution of human rights norms and related compliance pioneered by Thomas Risse, Stephen Ropp, and Kathryn Sikkink (1999, 2013), states are often the dominant actors. Recent scholarship on BHR highlights the role of companies as human rights norms protagonists.

But comparably little work explores the role that average community members play as respondents to and agents of the norms-making processes integral to the evolution of BHR ideas and processes (exceptions include Bauer 2011; Rodríguez-Garavito 2017a, 22–32; Meyersfeld 2017).

The same gaps are evident in the large volume of work (commonly referred to as gray literature) published by nongovernmental organizations (NGOs), government agencies, or consulting entities that have analyzed stakeholder dialogues. Much of this work is focused on individual case studies (e.g., Reese 2011) of lessons learned by corporations or by lead NGOs involved in dialogues. It is largely prescriptive in nature, instructing key parties on how to participate in dialogue rather than evaluating the intrinsic nature of participation.[5] Both the gray literature and academic scholarship on BHR have tended to focus on corporate incentives for entering into dialogues (Sherman 2009).

Comparably fewer studies have analyzed the role for community members in the dialogue process, or their normative understandings or incentive structures. An increasing number of authors note the gap (Wilson and Blackmore 2013, 11, cited in Knuckey and Jenkin 2015, 21; de Felice 2014; Kaufman and McDonnell 2015; Brown 2007; International Alert and Engineers Against Poverty, 2006). As recently as June 2017, a nongovernmental research team led by staff of MSI Integrity (i.e., Multistakeholder Initiative Integrity) concluded a two-year global assessment of 45 such consultative mechanisms by pointing to their overwhelming lack of provisions for meaningful community engagement or redress (Collins, Evans, Hung, and Katzenstein 2017, 3). "The fact that most

MSIs have not been driven by rights holders themselves," the report noted, "and that governing processes may not be adequately responsive to the perspectives of their most vulnerable stakeholders, calls into question their credibility and capacity to have positive local-level impacts" (Collins et al. 2017, 10). These authors call for qualitative research (Collins et al. 2017, 12) that "could also analyze how and to what extent communities desire to be involved in initiatives" (Collins et al. 2017, 10). This book takes up their call, embedding the research in a deeper theoretical and historical framework and grounding it with the perspectives of people at the grassroots.

My theoretical point of departure is remedy: in a general sense, remedy is defined as the "legal means to recover a right or to prevent or [to] obtain redress for a wrong" (*Merriam-Webster* 2017). Three aspects of this definition create challenges for economic rights fulfillment in the context of stakeholder dialogue. First, remedy is to be mediated through the legal system. But legal proceedings related to remedy, when they occur, have tended to pivot around violations of physical integrity rights (e.g., rape of women in the environs of gold-mining sites) rather than violations of economic rights (e.g., endangering rights of subsistence or health). Even in legal systems in which poor people can seek damages jointly for economic rights violations, the practical difficulty of amassing the evidence, paying for counsel, and sustaining a case over time proves difficult for many.[6]

Second, the nature of claim-making intrinsic to this definition of remedy is retrospective. Two of the three main actions that a claimant can take (i.e., to "recover a right" or to "obtain redress for a wrong") require the claimant to be able to demonstrate past harm. This is reactive, not proactive, claim-making. Remedy interpreted in this manner precludes forward-looking analysis of the types of rights fulfillment over time that would be necessary to ensure progressive realization of economic rights, which have an inherently dynamic character (Ramasastry 2015, 250, citing Wettstein 2012, 739).

Third, the definition's reference to actions that would enable a claimant to "prevent . . . a wrong" implies a relationship between claimant and grantor that is implicitly individualistic in nature. The direct link between a single violator, a specific victim, and the prevention of future harm precludes remedy for structural harms. Yet violations of economic rights are deeply structural in nature (Meyersfeld 2017). Audrey Chapman, originator of the "violations approach" to economic and social rights monitoring (1996), has since argued for a minimum core approach instead (2007) because it is dynamic in nature. Standards of fulfilment evolve over time apace with changes in the guarantor's resource level, which means that as resources increase, guarantors are held to a higher level of fulfilment, but when resources are constrained, claimants are also protected from rights "backsliding" (known as retrogression) by locking in a minimum floor of protection.

The theoretical challenges intrinsic to remedy just discussed are amply reflected in the Ruggie Principles themselves. These principles outline three main categories of mechanisms for ensuring remedy: state-based judicial remedies; state-based nonjudicial remedies; and non-state-based, nonjudicial remedies (UN Human Rights Council 2011, 28–31), including many of the types of MSIs discussed already. They also stipulate criteria for evaluating the effectiveness of nonjudicial grievance mechanisms in particular. Effective mechanisms should be legitimate, accessible, predictable, equitable, transparent, rights-compatible, and dialogue-based and should offer a source of learning, according to the Ruggie Principles (UN Human Rights Council 2011, 33–34). But the substantive content of these criteria is still in flux, and the approaches to remedy vary widely in practice, as do the roles of community members in designing and executing them. This ambiguity has given way to a variety of practical mechanisms for enacting whatever stakeholder dialogue might mean (Fransen and Kolk 2007). Many such mechanisms are characterized by an overarching tendency toward a top-down opening of space for dialogue into which community members must fit their grievances, and a constrained temporal focus—either backward-looking or preemptive, but not both.

The rules of the game (i.e., when and where the dialogue takes place, in what language) are typically set by companies, not by local community members. Companies (not communities) typically determine the time horizon for calculating the scope of past or future harms. Consequently, stakeholder dialogue as a framework for negotiating risk underestimates the depth of preexisting conflict over economic rights (i.e., does not look back far enough) and may fail to anticipate the challenge of ensuring progressive realization of economic rights over time (i.e., does not look far enough forward). If it is to contribute substantively to remedying economic rights violations that occur in the course of contemporary business practice, stakeholder dialogue will have to evolve to engage a wider range of people in local communities in setting the scope, terms, and goals of such interaction. This book explores the challenges inherent in doing so.

PLAN OF THE BOOK

The chapters that follow analyze stakeholder consultation over time, across regions and sectors, and from bottom up and top down. Chapter 2 takes an historical approach, developing a genealogy of state and corporate engagement with poor people from the 1980s to the present that reveals the continuity of constraints on their ability to fully claim their economic rights. By laying out the structural roots of underdevelopment and corresponding lack of attention to corporate responsibility for remedy in the early stages of accelerated global economic integration (from the late 1970s onward), this chapter demonstrates that the contemporary shortfalls of stakeholder consultation have much deeper roots. Chapter 3 maps the scope of contemporary corporate stakeholder engagement practices across multiple regions and industrial sectors by analyzing data on 7,000 companies in the public database of the Business and Human Rights Resource Centre (BHRRC). The chapter reveals the disproportionate concentration of dialogues in the extractive sector across multiple world regions. The chapter also lays out the field research methodology and case selection justification that informs the subsequent chapters, demonstrating why multiple methods of inquiry and multiple forms of data are necessary if we are to fully understand the potential and limits of contemporary stakeholder consultation.

INTRODUCTION | 13

Chapters 4 and 5 use a comparative case study framework to analyze the complex local communities in which global textile manufacturing takes place and the factors that influence the prospects for stakeholder consultation. The settings are two manufacturing towns in the Dominican Republic where collegiate apparel is produced, specifically the T-shirts, sweatshirts, and other textiles ultimately embellished with college logos (including that of my own university). This is a segment of the apparel market with high consumer attention to corporate responsibility, and the companies in the two towns selected for comparison have employed, respectively, a WSR factory governance model and a BHR-driven mode of supply chain management and community engagement.

Drawing on original data from 43 original grassroots-level interviews conducted in the two manufacturing communities in summer 2017, the "method of difference" (Mill 1843) comparative case study design limits variation on key factors (e.g., industrial sector, geographic region, export market) in order to explore variation in public attitudes regarding stakeholder dialogue. Listening to people at the grassroots level thus illuminates the potential and limits of WSR and BHR strategies, the structural roots of poverty, and the inherent complexity of poor communities that render the prospects for remedy more challenging than many proponents of stakeholder dialogue suppose.

Chapter 6 assesses the prospects for policy reform, both in relation to BHR-led MSIs and WSR models. To explore the potential and constraints central to both models, this chapter draws on original data collected in 2017 at an international conference hosted by the University of Connecticut Business and Human Rights Initiative, involving academic,

NGO, labor union, and business representatives from across the United States, Europe, Latin America, and Asia. The chapter also draws on theories of innovation in failure-prone settings (from systems engineering) and on practical examples of creative approaches to managing supply chains and strengthening social and economic rights promotion in multiple settings and across industries (particularly in the WSR arena). The book concludes with a renewed call for more robust theory and practice on forging economic rights remedies that are inclusive, dynamic, and adaptive to the ongoing challenges unfolding in supply chains that link stakeholders in ever more complex and challenging ways globally.

CHAPTER 2

A Genealogy of Community Consultation

RECENT INNOVATION IN BUSINESS AND human rights (BHR) theory and practice—in particular, the emergence of the norm of corporate responsibility to "remedy" abuse and the norm of community entitlement to remedy—has not taken place in a vacuum. Rather, these norms have emerged as part of a deeper genealogy of popular participation in economic policymaking that undergirds contemporary stakeholder dialogue. Drawing on a combination of primary source documents from nongovernmental organizations (NGOs) and intergovernmental bodies along with existing academic scholarship, this chapter identifies three main phases in this genealogy. The first is a phase in which corporate actors engaged local community members as a form of damage control (circa 1980s). The second is a phase in which participation took place through victim testimonial (1990s). Third is the present phase, in which participation has emerged as vehicle for entitlement (2000s). At each juncture, specific new mechanisms have emerged that enlarge and simultaneously constrain the nature of stakeholder involvement in economic rights policymaking.

Foucault's concept of genealogy (1980, 82) offers a vehicle for excavating subordinated ideas and in the process empowering subaltern people who were the originators of such ideas—both by demonstrating the value of their contributions and by shedding light on the exploitation that has led to the burying of their ideas in the first place. This chapter creates a genealogy aimed at illuminating the deeper limits of community involvement in stakeholder dialogues. It explores how grassroots actors have faced constraints in processes of consultation and participation ostensibly aimed at including them in order to protect their economic rights.[1] Three critical junctures are key to understanding the genealogy of subordinated participation, along with the specific forms that enlarge and simultaneously constrain participation at each juncture.

PARTICIPATION AS DAMAGE CONTROL: THE 1980S

The legacy of ritualized participation in economic policymaking and implementation runs deep. This chapter does not extend back to the colonial period, during which colonial powers employed local people in positions of relative authority vis-à-vis others in society in order to foster control and reinforce coercive extraction.[2] Rather, the chronology here begins following the petro-boom of the 1970s, which financed a wave of public and private sector borrowing by governments throughout the developing world and led ultimately to a global debt crisis by the 1980s.[3] Throughout the debt crisis, governments implemented structural adjustment policies (SAPs) aimed at reining in spending, reordering government accounts, and stabilizing the financial sector. Access to financing through public international financial institutions such as the International Monetary Fund (IMF) and World Bank was conditioned on adopting this policy package.

The initial formulation of SAPS happened with little long-term consideration for the social effects of extensive retrenchment in social welfare policy (Abouharb and Cingranelli 2008), and as the human cost of deep cuts in social welfare spending became apparent, the United Nations Children's Fund (UNICEF) sounded the alarm. UNICEF published its 1987 landmark study *Adjustment with a Human Face* (Cornia, Jolly, and Stewart 1987) toward the end of the "lost decade" of the 1980s. The main report analyzed the effect of macroeconomic adjustment on children and women's well-being across Asia, Africa, and Latin America. Subsequently, region-specific studies emerged, including the 1979 volume *The Invisible Adjustment* (produced by Latin American scholars and policymakers under the direction of UNICEF's regional field office in Santiago, Chile). *The Invisible Adjustment* captured the wide range of ways in which (quoting the editors) the then-"present economic crisis of social disinvestment is being financed principally from the resources of a 'social fund' provided by the superhuman efforts of poor women" (Rocha, Bustelo, López, and Zúñiga 1979, 12).

Social scientists pivotal in the production of the *Adjustment* reports—including British economist Richard Jolly—would ultimately become leaders in the creation of the UN Development Programme's *Human Development Report* in 1990 and, with it, the new "Human Development Index" (HDI), which integrated measures of income, longevity, and education into a single indicator by which countries could be ranked. The hegemony of the HDI and other development indicators has already been deeply critiqued by contributors to volumes edited by Merry, Davis, and Kingsbury (2015), by Rottenburg, Merry, Park, and Mugler (2015), and by Cooley and Snyder (2015). Rather than revisit their arguments here, this chapter takes as its point of departure the connection between the rise of efforts to bring to light the "superhuman efforts of poor women" through this reporting and the subsequent challenge of bounded participation in economic policymaking.

Following the exposure of the human cost of SAPs, both the IMF and World Bank made efforts to adapt this policy framework to address the effects of across-the-board cuts in formerly subsidized or wholly government-provided health, education, nutrition, and other social programs. Subsequently, the IMF and World Bank introduced a process of community involvement in the drafting of Poverty Reduction Strategy Papers (PRSPs) in 1999.[4] The PRSP process aimed at increasing popular buy-in for macroeconomic policy change by eliciting citizen input into policy design. Representatives of government social ministries throughout the countries involved in SAPs enlisted members of communities in discussions of development frameworks through the PRSP process. But this mechanism has remained a source of considerable academic debate. Critics charge that the involvement of civil society is overly scripted through the PRSP process in order to weed out dissenters and to neuter opposition and/or tame it through inclusion (Lazarus 2008; Dijkstra 2011; Action Aid 2004). By contrast, supporters consider the strategy of inclusion pivotal to the success of adjustment (World Bank 2004).

GENEALOGY OF COMMUNITY CONSULTATION | 19

Running parallel to these top-down efforts at increasing the scope of development policy to include grassroots-level impact analysis, a series of high-profile industrial disasters occurred in the 1980s and early 1990s that generated corresponding popular pressure to address the rights of victims. A timeline compiled in 2015 by the UN Global Compact (a membership forum of over 10,000 companies that annually disclose data on their efforts on corporate social responsibility) includes pivotal events in the "emergence of the modern corporate sustainability movement" (United Nations Global Compact 2015, 31–34; for background on corporate social responsibility, see Vogel 2006). Among them are the 1977 consumer boycott of Nestlé in response to the marketing of baby formula in the developing world; the 1989 *Exxon Valdez* oil spill; the 1993 assassination of Kenyan environmental and indigenous rights activist Ken Saro-Wiwa in the wake of protests against Shell Oil in that country; and the 1995 Brent Spar oil platform protests in the United Kingdom.

But the 1984 explosion in rural India at a pesticide manufacturing facility operated by the US multinational corporation Union Carbide stands out for its scope and for the types of longstanding popular mobilization it has engendered. The Bhopal disaster exposed thousands of rural people to harmful gasses and killed between 2,200 and 3,700 people in the environs of this community in the state of Madhya Pradesh. According to the UN Global Compact (2015, 31), Bhopal is considered "the world's worst industrial disaster," with a final death count that remains contested and ongoing health complications for survivors (Eckerman 2005; Ruggie 2013, 6–9; Deva 2012, 24–45). Seeking unsuccessfully to defend itself against criminal and civil suits in the Indian courts and against prosecution under the US Alien Tort Claims statute, Union Carbide remains a target of legal action.

A relatively under-explored aspect of the Bhopal experience has been the community-based networking it has spawned across borders. As Joanne Bauer explains, community-based organizations in both India and the United States have worked with survivors of the Bhopal disaster to foster linkages between Indian and US-based groups similarly affected by chemical exposure in the wake of industrial accidents (Bauer 2011, 191–192). In part because poor people do not have standing as a "protected class" in US law, activists in West Virginia (where chemical storage facilities house the same compounds that devastated Bhopal) have placed their struggles for safe regulation of hazards in a broader historical context, thus forging a connection between the two communities in the interest of pressing industry and government into more effective regulation.

These efforts are mirrored in other settings in the United States, where activists have engaged in creative "venue-shopping" in order to bind the power of corporate actors while raising community and broader public awareness of the right to remedy. US-based advocates in minority communities in Louisiana and among the Western Shoshone tribes in Nevada, for example, have demonstrated the racially disproportionate effect of industrial harm on minority communities by employing a multipronged strategy of "shaming and blaming" both the US government and US multinationals for ongoing ineffectiveness in remedying environmental racism (Bauer 2011, 185–191). They have used reporting procedures under the International Convention on the Elimination of All Forms of Racial Discrimination (ICERD)—which the United States has ratified—along with the InterAmerican Commission for Human Rights to spotlight the US government's ineffectiveness in regulating the racially disproportionate effects of corporate pollution and to emphasize the joint responsibility of government and corporations to safeguard communities' economic and cultural rights. All of these examples offer evidence of strategic grassroots action aimed at transcending the limits of participation as damage control.

PARTICIPATION AS TESTIMONIAL: THE 1990S

Global manufacturing and trade increased in scope and depth in the 1990s in ways unparalleled historically. According to Ruggie, fully half of all world trade became composed of "internal" transactions between "networks of related corporate entities" (2013, xv) during this period, and states significantly reformed regulatory structures in an effort to facilitate increased financial and trade flows.[5] The end of the Cold War gave way to the movement "offshore" of manufacturing and service jobs, facilitated in part by trade integration channeled initially through the General Agreement on Tariffs and Trade (GATT) and later through regional trade agreements such as the North American Free Trade Agreement (NAFTA) as well as the eventual creation in 1994 of the World Trade Organization.

While states encouraged the shift toward global production through domestic regulatory reforms, firms themselves took advantage of changes in technology that made doing business internationally much easier than in the past. The 1990s marked the emergence of innovations in communications and transportation technologies, including the rise of the Internet and with it computerized banking along with the development of barcode inventory and electronically networked logistics management, the expansion of deep sea ports for containerized shipping, and decreases in the cost of commercial airline flights—all of which combined to make it easier and cheaper for US- and European-based multinationals to extend their supply chains, marketing, and distribution internationally (Gereffi, Humphrey, and Sturgeon 2005).

These same technological changes paved the way for the ratcheting up of consumer demand for lower prices and faster turnaround times linked to the advent of "fast fashion," with its emphasis on highly trend-driven and seasonally driven production of low-cost garments (Dickson, Loker, and Eckman 2009; Rivoli 2014; Anthony and Taplin 2017). During the same decade, industrialized country–based importers pushed to consolidate their supply chains among "preferred" suppliers (Schrank 2013, 423) in part because they could manage inventory in a more streamlined manner owing to changes in technology and in part because consumers in industrialized countries demanded greater corporate accountability for human rights conditions of workers involved in garment production in developing ones.

Indeed, the expansion of the 24-hour news cycle during the 1990s gave way to a resurgence of public interest in labor rights issues more generally because video, audio, and still photographs could be broadcast readily, first into the living rooms and later onto the personal computers (and ultimately, by the early 2000s, onto mobile phones) of consumers in the industrialized world, where finished products were sold (Winston and Pollock 2016). While landmark social change photographers such as Dorothea Lange and journalists such as Edward R. Murrow had chronicled abuse of labor and economic rights in manufacturing and agriculture throughout the United States during the 20th century, the speed with which photographic images and video could be transmitted from the mid-1990s onward accelerated the trend toward visually driven human rights advocacy (Ristovska 2016; Brysk 2013, 132–162).

Images of labor and economic rights violations, in particular, were central to the work of photographers such as Sebastião Salgado in Brazil and Fernando Moleres in Spain, who partnered with NGOs and UN agencies—including the International Labour Organization (Moleres 2000)—in awareness-raising campaigns aimed at consumers, firms, and governments. The 1990s birthed a new wave of testimonial photography and filmmaking married with direct advocacy by human rights NGOs that was unprecedented in its reach and scope (Pruce and Budabin 2016). For example, child labor–focused photographer U. Roberto ("Robin") Romano and his collaborators filmmakers Len and Georgia Morris, along with photographer Sara Terry (whose work focuses on war and its aftermath), developed projects in conjunction with national and international networks of advocates whose visual stories became central to consumer campaigns, legislative reform agendas, public outreach, and education.[6]

Nongovernmental and advocacy organizations—some like the century-old US National Consumers League, others relative newcomers like the Netherlands-based

Clean Clothes Campaign—developed computerized lists of people to alert about contemporary sweatshops and to mobilize around policy reform efforts. They used online activism strategies (such as petitions) while at the same time resurrecting strategies like consumer boycotts and use of the Alien Tort Claims Act[7] to carry out popular protests and legal action in defense of people harmed by corporate activities, wherever in the world they took place. At the same time, activists developed new tactics like public "die-ins," in which activists feigned their deaths in front of corporate headquarters, or attempted to take over shareholder meetings of publicly traded companies (long a focus of ethical investing since the 1970s)—all while the cameras rolled and the archival footage could be stored on NGO websites (or, by 2005, on public hosting sites such as YouTube). Advocacy organizations such as WITNESS emerged in the late 1980s, formalizing the use of handheld video as a source of awareness-raising and legal evidence-gathering (Ristovska 2016; McLagan 2003).

Workers who produce collegiate apparel became a particular focus of social solidarity efforts on college campuses throughout the United States (Elkins and Hertel 2011; Adler-Milstein and Kline 2017, 23–26, 82–90, 141–151). When labor rights advocacy in the textile sector took place from the middle to late 1990s, it was often brokered by peak nongovernmental monitoring organizations based in the United States, such as the Fair Labor Association (FLA) and the Workers' Rights Consortium (WRC), founded respectively in 1996 and 2000. These nongovernmental, membership-based organizations emerged in order to enforce compliance with labor rights standards in supply chain management, principally in light industry such as textile manufacturing. Although commercial auditing and certification organizations have been involved in the burgeoning compliance industry (see Locke 2013; O'Rourke 2003; Hertel 2010), the normative "heart" of the labor rights compliance and monitoring sphere has been NGO-driven monitoring efforts, spurred by advocacy groups that have worked with issue- and/or region-specific networks to safeguard labor rights.

Worker testimonials were part of a wave of citizen involvement in public "tribunals" organized by NGOs throughout the 1990s. In some ways, they paralleled in form the "truth commissions" emerging in the wake of transitions to democracy throughout the developing world following the Cold War (Hayner 2010) because they sought to expose wrongdoers and create a public record of past harms. But the organizers of worker tribunals often aimed to do more than create a public record of the past. They sought to make visible the people involved in global supply chains and to insist that seeing required action on the part of companies, states, and consumers, each in their own sphere—through changes in pricing and pay scales, reinvigorated labor rights enforcement, and ethical consumption.

This process of truth-telling, public reckoning, and demanding played out both in NGO-organized tribunals and in formal settings such as major UN conferences hosted over the course of the decade. At each conference, NGOs orbited formal state delegations and hosted activities in their parallel "NGO Forums" aimed at mutual education and coalition-building (Clark, Friedman, and Hochstetler 1998). The UN system, which had previously restricted nongovernmental participation to an elite set of organizations with "consultative status" that had been active in the UN debates since its founding, was forced to accommodate a burgeoning number of networks and groups that sought to participate in the major conference organized annually in the 1990s (i.e., the 1992 UN Conference on Environment and Development at Rio de Janeiro; the 1993 World Conference on Human Rights at Vienna; the 1994 UN International Conference on Population and Development at Cairo; the 1995 World Summit on Social Development at Copenhagen; the World Women's Conference at Beijing in that same year; and the 1996 Second UN Conference on Human Settlements, or "Habitat II," held at Istanbul).

The UN created a special "roster" status for groups outside the select "consultative" grouping. These groups registered to participate in the parallel UN Forums. While excluded from formally introducing commentary into the process of drafting under way among UN country delegations, these "roster" status groups nevertheless were lively participants in regionally based caucuses that supplied members of country delegations and "consultative status" NGOs with preferred language to introduce into the evolving soft-law "Declaration and Programme of Action" created at each conference.

Academic literature on the resurgent power of norms in international relations piggybacked on this real-world activism, and a range of scholars sought to sort out the direction, intent, and impact of the "shame-and-blame" tactics central to human rights advocacy, many focusing initially on civil and political rights–dominated action (Keck and Sikkink 1998; Risse, Ropp, and Sikkink 1999). An increasing number have focused as well on advocacy around economic and labor rights (Hertel 2006; Brooks

2007; Gray and Hertel 2009; Chong 2010; Nelson and Dorsey 2008; Baer 2015). The persistent challenge in all of this scholarship on victim-driven advocacy campaigns has been to make visible the structural factors that undergird oppression—including the complicity of citizens and consumers in industrialized countries (see Pogge 2008).

For example, gendered protests in the context of trade integration reveal multiple power dynamics. Hertel's analysis of Mexican tribunals on reconciling maternity and work held in the mid-1990s reveal hierarchies among activists at the subnational as well as transnational levels (2006, 55–85). During these forums, factory workers, public school teachers, bank employees, and academics alike bore public witness to widespread and systematic workplace discrimination against pregnant workers in Mexico. However, women involved in campaigns where their testimony was vital to "shaming and blaming" state as well as corporate actors often noted the internal hierarchies within and among advocacy networks that this form glossed over. And they criticized the lack of concrete policy reforms in the wake of testifying. The catharsis of having their say was often short-lived, while the longer-lasting structural reality of low-wage labor and the ongoing double burden of paid and unpaid labor persisted.

Yet the rise of popular protest on labor and economic rights in the "sweatshop decade" of the 1990s meant that public or private institutions that resisted engagement with NGOs increasingly did so at their own peril. For example, the 1998 "Battle at Seattle" waged by a coalition of labor and environmental NGOs sought to push the World Trade Organization to open its policy process to accommodate alternative ideas of how global trade should be regulated

(Hertel 2005). Extensive street protests at the WTO's Third Ministerial meeting in Seattle ultimately prevented trade negotiators from even getting to the conference venue from their hotels and completely stalled the conference proceedings, which ended without a final communique or significant advancement in the negotiations. The WTO subsequently created vehicles for NGO comment and began to post online many of the key documents integral to upcoming trade negotiations. Tellingly, however, the ministerial meeting immediately following Seattle was held in Qatar—not a state with widespread freedom of movement or expression but instead a location chosen, critics argued, in order to limit popular protest around trade policy (Human Rights Watch 2001).

This tendency toward circumscribed participation deepened in the trade arena throughout the 1990s. Even where engagement of non-state actors did take place and even when NGOs themselves mobilized such participation, the challenge remained of ensuring that the people central to the process were protected. The act of offering testimony as a vehicle for raising awareness, gathering evidence, or pressuring companies or governments to negotiate could potentially retraumatize or endanger the very people such strategies were intended to help. Organizing tribunals could reinforce multiple and often intersecting hierarchies among activists and those they sought to help. Or it could create the illusion of progress but leave the structures of power untouched (Brooks 2007).

Throughout the 1990s, as groups such as the National Labor Committee and Equal Exchange led tours of victims throughout the United States and Europe, US and European consumers had their eyes opened—but at times at the expense of people involved in testimonial tours, who faced the risk of significant repercussions upon returning home. Who "owned" their stories, or their victories—the storytellers themselves, or the NGOs who had organized and funded their trips and coached them in delivering their remarks? These are challenges central to the victim-narrative advocacy genre (Meyers 2016) that underpins this second phase in the genealogy of participation and consultation.

PARTICIPATION AS VEHICLE FOR ENTITLEMENT: THE 2000S AND BEYOND

Embedding the emergence of multi-stakeholder consultation within the genealogy of participation exposes the inherited limitations both on individual agency and on structural reform that are present in current forms of stakeholder consultation. The early 2000s opened a new set of frameworks for participation, beginning with the creation of the UN Global Compact itself in 2000, and followed by the creation of the post of UN Special Representative on Business and Human Rights in 2004 held by John Ruggie.

Starting in 2004 and extending through two terms (ending in 2011), Ruggie met with literally thousands of people in communities affected by business conduct across the world. He also made a point of engaging representatives of businesses themselves, large and small and in multiple industries, in crafting a set of principles designed to govern stakeholder relations in the modern era through a protect/respect/remedy framework in which government would set an enabling environment (through laws, enforcement action, and macro-policy design) for *protection* of human rights; business would *respect* laws as a minimum floor; and government and business would be *responsible jointly for remedy* for those people who suffer abuses in the context of business activity. The third plank of the UN Principles on Business and Human Rights (or "Ruggie Principles") has become the thorniest.

The twin evolution of the Global Compact as a forum and the Ruggie Principles as a soft-law framework for shaping remedy has taken place against the backdrop of global economic crisis (in 2007), the resulting food crisis (2007), and a rising tide of popular discontent with economic inequality that burgeoned locally in the "Occupy Wall Street" movement (2011) and was amply reflected in the 2015 US presidential race (Malone, Nayak, Bolton, and Welty 2013). A major challenge in this context is to define remedies that adequately address the underpinnings of power inequalities at the heart of businesses' violations of human rights. A critical first step is changing the terms of popular participation in the "consultations" emerging through this BHR framework. At present, stakeholder consultation appears to offer far more benefits to companies (i.e., in image maintenance and constraining demands) than to communities.

This is not surprising from either a theoretical or practical perspective. Former World Bank staff member and development theorist Deepa Narayan has argued that amidst the emergence of interdisciplinary scholarship on empowerment in the early 2000s, "work has only recently begun on construction of an analytical framework on empowerment that can be used to guide *state reform and action*" (2005, 4, emphasis added). Chastened by the often negative response to PRSPs, Narayan and others in her World Bank research team began to theorize how to increase empowerment in the development process—but the emphasis was largely on state–civil society relationships, not on the private sector's role in fostering (or inhibiting) empowerment. Nor did it interrogate the structural forms of inequality inherent in global economic relations.

A new wave of literature produced on supply chain management, written largely by scholars of business and business ethics, focuses largely on the challenges faced by Western firms that contract production abroad, rather than their subcontracting partners based in developing countries. Community relationships are almost entirely left out of these analyses. Richard Locke (2013) makes a compelling business case for paying greater attention to the constraints faced by subcontractors in the global South. Through extensive field-based research in multiple regions over multiple years, coupled with analysis of thousands of factory auditing reports from multinationals that globally subcontract production or manufacture themselves, Locke has demonstrated that punitive sanctions and short-term contracting relationships fail to create incentives for upholding labor and environmental rights. If the goal is better conditions on the factory floor in developing countries, he argues, we cannot ignore the challenges that developing country producers face. Nor can we ignore the larger governance gaps in those same countries and the lack of capacity and/or political will to address the root causes of labor rights abuse (e.g., poverty, corruption).

Yet as path-breaking as Locke's work has been in the field of supply chain management and regulatory theory, it leaves to the sidelines the question of community involvement in stakeholder consultation. Soundararajan and Brown (2016) acknowledge that much of the literature on supply chain management and stakeholder consultation still focuses overly on the interests of Northern consumers and multinationals. The mechanisms for stakeholder consultation are still lopsidedly focused on building value for Northern business partners in the supply chain.

In an effort to fill the gap, Soundararajan and Brown are carrying out original empirical work with Indian subcontractors involved in export-oriented light manufacturing, but community members remain largely peripheral to their analysis, regarded as "derivative" stakeholders (2016, 99). The authors acknowledge the yawning governance gap at the national level in many developing countries, along with broader questions of international distributive justice (2016, 86) and "fairness-based" concerns (2016, 88). But their template for enhancing stakeholder involvement in voluntary governance of global supply chains leaves out discussion of vehicles for community consultation, redress, or involvement in institutional redesign.

Soundararajan, Brown, and Wicks identify "institutional voids" (2016, 5) as a challenge to managing supply chains and acknowledge that "[p]owerful businesses thus assume the role of a void-filler, using lobbying mechanisms to gain power to realize their interests in the absence of institutional infrastructure... [Multi-stakeholder initiatives] are one of the mechanisms that global businesses use to fill the void" (2016, 9). Here the authors focus on increasing stakeholder value—but only among business partners in the supply chain, thus leaving the interests of community members outside the scope of their analysis and proposed reforms. They acknowledge that if some actors are marginalized in the stakeholder dialogue process and "feel like they have no voice, or meaningful involvement ... they will be more inclined to either ignore it or seek to undermine it" (2016, 18).

The challenge of sustaining civil society involvement in local activities of the UN Global Compact is a case in point. Its membership includes companies of varying sizes, governments, nongovernmental organizations, and academic institutions, and is growing overall (United Nations Global Compact 2011, 22) but the direct involvement of NGOs is dropping. As the Global Compact itself reports, since 2009 civil society participation in its local networking processes has dropped from an average of 60% to 26% of those involved (United Nations Global Compact 2011, 31).

A 2015 survey conducted by the Boston Consulting Group and MIT Sloan School Management Review[8] echoes these findings, revealing a clear trend toward greater numbers of stakeholder consultations happening worldwide, though they are concentrated (over 60%) on labor issues—often mediated vis-à-vis formal labor unions[9]—and touch less upon either general human rights or corruption issues (United Nations Global Compact 2015, 98). Although over 2,100 civil society organizations have registered as participants in the Global Compact, "only a small fraction of the participating organizations contribute actively" (2015, 117). Rather than question why popular participation is down (might the groups simply see it as window-dressing and/or have become reluctant to legitimize this forum by participating?), the Global Compact office has responded by creating a "Communication on Engagement requirement for non-business participants in 2013" and taking a punitive stance: "Failure to communicate will . . . result in delisting from the Global Compact database. The expectation is that this may significantly reduce the number of civil society organizations participating in the next year" (2015, 117), which the Global Compact neither problematizes nor seeks to address by acknowledging the profoundly structural power imbalances among Global Compact members themselves.

Even the literature on the cutting edge of the supply chain field skirts the questions of who pays for remedy and broader issues of redistributive justice and community control over resources. It assumes that the act of participation—the ritual of being consulted, the scripted role for grassroots communities in the BHR conversation—is itself intrinsically important enough to satisfy the grievances of people who suffer in the course of business doing business in their communities. Participation on these terms assumes that grassroots people have the time and disposable resources to take a seat at the table and leave their other responsibilities aside. Uncompensated participation itself is the focus of critique by anthropologists and sociologists interested in local participation in human rights policymaking and processes (Buerger and Holzer 2015).

Contemporary stakeholder dialogue as currently practiced thus often leaves communities as bystanders (or largely ceremonial participants) *by design*. Building trust through participation, however, cannot happen in the absence of meaningful resource-sharing and credible sanctioning. Soundararajan, Brown, and Wicks invoke Amartya Sen's notion of participatory governance in arguing that "long-term sustainable implementation" of consultation requires that companies, suppliers, the state, and the civil society together engage in "capability development through a localized participatory approach" that enables "stakeholders to *'lead the kind of lives they value and have reason to value'*" (2016, 31, quoting Sen 1999, 18, italics added).

Exposing the genealogy of constraints integral to contemporary stakeholder consultation is but the first step. Challenging the power hierarchies they reflect also hinges on building the capacity and resources of community members to define the kind of lives they value and forging the skills and alliances to defend them, as the case studies and policy analysis in the successive chapters demonstrate.

CHAPTER | **3**

Global Landscape and Local Contexts for Stakeholder Consultation

TO UNDERSTAND THE LIMITATIONS AND potential of stakeholder engagement as a tool for remedy in the business and human rights (BHR) arena, we need more than the historical account laid out in the previous chapter. We also need a contemporary map of the industry sectors and geographic regions where company–community consultations are currently taking place, along with grassroots evidence from local people themselves about their perceptions of the prospects for stakeholder dialogue and remedy in their communities. This book thus works at several levels of analysis: exploring global trends in this chapter, which sets the stage for the highly localized comparative case study chapters that follow it, then concluding with a policy implications chapter that moves between the two levels. Different types of data are necessary at each turn.

MAPPING TRENDS IN THE SPREAD OF STAKEHOLDER DIALOGUE

To identify regional and sectoral trends that mark the spread of this form of BHR practice, this chapter analyzes primary and secondary data collected by nongovernmental organizations (NGOs) in the BHR arena. Rigorous, comparable data on corporate BHR practices are scarce in general, data on community consultation and/or remedy even more so.

There are not global reporting protocols incumbent on all companies in relation to BHR, and national standards for reporting differ significantly. Environmental sustainability reporting standards tend to be more robust and more widely and uniformly adopted by companies than social reporting standards (O'Connor and Labowitz 2017, 10).[1] The NGOs that conduct research on BHR cull data released by companies and supplement it with news accounts from media sources as well as reporting from governments, international organizations, local advocacy or labor organizations, and individual eyewitness accounts. Therefore, the information compiled is inherently partial in nature and may be biased toward large companies or English-language media or "herding" effects in reporting, or the type of data that companies choose to release, or differences in regulation by state or region or industry.

The global or industry trends discussed in this chapter are thus necessarily subject to the limitation of the principal data source analyzed here—specifically, the data archived by the Business and Human Rights Resource Centre (BHRRC), a nongovernmental research organization that monitors BHR activities of 7,000 corporations worldwide. Data produced by other research organizations and analyzed in this chapter for comparative purposes with the BHRRC data should be similarly qualified.[2] Data gaps necessarily constrain generalizations about the scope or depth of related practices.

With these caveats in mind, this chapter offers an overview of salient trends derived from analysis of BHRRC data. Founded in 2002 with offices in London and New York and researchers based in 14 countries worldwide, the BHRRC is a key repository of data compiled by NGOs, unions, state and international agencies, and academics worldwide. The BHRRC archives the data it collects by company, issue, and country in a database that is fully searchable, free, and open to the public. Also archived are corporate responses to allegations of corporate involvement in human rights abuse along with data produced by firms themselves on their efforts to promote rights. Reports and responses are treated equally as searchable pieces of data within the BHRRC data portal.[3]

The BHRRC does not convene stakeholder dialogues itself but instead acts as a key repository of information about them. It also served as the main nongovernmental website at which UN Special Representative for Business and Human Rights John Ruggie posted all of his working papers and official reports over the six-year course of his UN mandate. Hence, it is a significant source of publicly available data that can be used to analyze the emergence of varying modes of stakeholder dialogue across industries and regions worldwide.

From the BHRRC's inception in 2002 until March 2018, analysis of BHRRC data had to be carried out manually by logging onto its website and inputting search terms from a predetermined menu of categories formulated by the BHRRC for use in searching the database. This system was suitable principally for hand-coding, and the initial research integral to this chapter was carried out in this manner in 2015 and 2016.[4] Beginning in early 2018, however, the BHRRC began developing an application programming interface (API) for its website through which the archived data could be accessed more efficiently. This chapter features initial findings derived from this new presentation of the BHRRC data.[5]

ORIGINAL DATA ANALYSIS, METHODS, AND FINDINGS

The earliest version of the API[6] enables users to query the entire database and receive information in standard JavaScript Object Notation (JSON) format. Using a programming language/software environment of choice, users can, in seconds, systematically sift through and precisely analyze thousands of articles, reports, web references, company responses, and other documents. At the time of this writing, the API is available only as a private beta, although anyone can register and request access; plans are in place to make the API publicly available.

We used the API to obtain paginated lists of all existing "Stories" at the time of coding and all possible "Categories" on the BHRRC's website. "Stories" can be a *standalone* news article or report, or a *collection* of related articles, reports, statements, and/or web references. "Categories" include countries, regions, sectors, principles, organizations, and company policy steps. Our goal was to search through and filter "Stories" based on their "Categories" in order to identify trends in how frequently key combinations of "Categories" appear together; the broader goal was to map trends in industry and regional prevalence of varying forms of stakeholder engagement. Through the API,[7] we obtained an initial total of 11,785 "Stories" and 776 "Categories," which we then narrowed to 11,150 stories in order to avoid the risk of over-counting items that were repeated within embedded collections.[8]

For a standalone "Story," it was relatively simple to manipulate the JSON file that we retrieved through the API in order to extract its "Categories." We chose from the BHRRC's 776 possible "Categories" our list of countries, industries, and mechanisms. Our final list of "Categories" contained 189 countries,[9] 11 industries selected based on the extensive nature of their global supply chains (i.e., including the extractive industries of *energy, diamond, metals and steel, mining, oil, gas and coal,* and *stone quarries* and the light manufacturing industries of *clothing and textile, footwear,* and *leather and tanneries,* as well as *water companies* and *pharmaceuticals*), and three mechanisms associated in the literature with stakeholder consultation (*complaints mechanism; free, prior and informed consent; impact assessment*). We then searched through our 11,150 "Category" sets to determine how many times each of the 6,237 possible combinations[10] of country, industry, and mechanism appeared together on the BHRRC's database. We extended this process to analyze combinations of region,[11] industry, and mechanism by adding up the number of country, industry, and mechanism references corresponding to all of the countries in each region.

Our analysis of the data captures industry-wide as well as regional tendencies in the BHRRC data that corroborate trends we have identified in the secondary literature (discussed later in this chapter) as well as qualitative observations made in the case study chapters that follow this one. Notably, the comparisons discussed later in the chapter focus on inter- and intra-regional trends among developing country world regions (Africa, Asia, and Latin America) because many of the references to North American or European countries in the BHRRC data relate purely to the headquarters of a given company, not the host country.

First, stakeholder dialogue processes (as captured in the BHRRC's three categories of complaint mechanisms; free, prior and informed consent; and impact assessments) are disproportionately concentrated in the extractive sector over the light manufacturing sector by a ratio of roughly nine to one in the three developing country regions (Africa, Asia, and Latin America). These ratios are even more pronounced within Africa and Latin America, where the corresponding ratios are 15 to 1 in both regions. In Asia, the ratio of consultation in extractive versus light industry is roughly five to one, meaning that considerably more consultation is happening in the light industry relative to the extractive sector in Asia than in the other two regions. Within all three regions, consultation numbers appear to be driven by reports of complaint mechanisms more so than reports of free, prior and informed consent or impact assessment. Overall numbers of reported consultation in pharmaceuticals and in water industries are considerably lower than either extractives or light manufacturing: together they report but one-quarter of the total number of consultations in light industry and 36 times less than that of extractives (Table 3.1).

Second, within the two major sectors of extractives and light manufacturing and across the three developing country world regions, a small number of countries typically account for a disproportionate share of the stories reported by BHRRC. Stories on consultation in light manufacturing in Asia are concentrated among four leading exporting countries, China, India, Indonesia, and

Table 3.1

TRENDS FROM BHRRC DATA				
	Africa	Asia	Latin America	Total
Light industry	41	95	30	166
CM	24	70	24	118
FPIC	6	10	0	16
IA	11	15	6	32
Extractive	613	442	460	1515
CM	110	92	71	273
FPIC	207	155	252	614
IA	296	195	137	628
Pharmaceutical	5	3	0	8
CM	0	0	0	0
FPIC	5	2	0	7
IA	0	1	0	1
Water	16	10	8	34
CM	2	0	0	2
FPIC	3	2	3	8
IA	11	8	5	24

Source: Business and Human Rights Resource Centre data, as of May 2018
CM, complaints mechanism; FPIC, free, prior and informed consent; IA, impact assessment

Bangladesh, which together account for 64% of all BHRRC references to light manufacturing and consultation in that region. The percentage is similar in Africa: 52% of the references on consultation in light manufacturing are accounted for by four countries: the Democratic Republic of the Congo (DRC), South Africa, Nigeria, and the Ivory Coast. In Latin American light manufacturing, reports on consultation cluster again among Brazil, Mexico, Colombia, and El Salvador (57%). The patterns of concentration among countries where stories of consultation are reported are similar in the extractive sector, region by region, although the countries vary slightly in each. In Asia, it is again India and Indonesia, but also the Philippines and Myanmar, that account for 55% of all references in the extractive sector. In Africa, six countries account for 51% of the references: Kenya, the DRC, South Africa, Nigeria, Tanzania, and Uganda. In Latin America, Peru, Colombia, Brazil, and Guatemala account for 56% of all references.

The overall number of counts is relatively small within the universe of 11,150 units in the data we analyzed. In the African extractive sector, 613 references are counted versus 41 in the African light manufacturing sector. In Latin America's extractive sector, there are 460 references in comparison to 30 in light manufacturing. As noted above, in Asia the ratio of extractive to light manufacturing is closer to five to one (i.e., 442 references vs. 95). In part, these relatively small numbers are a function of the precision and exhaustiveness of the search criteria we constructed for working with the BHRRC data. The numbers also reflect how human rights advocacy is driven in part by media reporting (Winston and Pollock 2016) around high-profile cases of company-specific abuse (i.e., news items that are picked up and cycle through news outlets) and in part by the relative strength of NGO networks at the grassroots level in different country and/or regional contexts (Hertel 2006).

To explain these trends, consider the literature on how variation in industrial sector (Mosley and Uno 2007) and supply chain management approaches (Locke 2013) affects variation in labor rights protection. The light manufacturing sector, with its flexible labor force composition and high factor mobility, experiences less stakeholder dialogue than the extractive sector, where sunk costs are significant and resources are not highly mobile. Firms in industries with more "flexible" labor structures and high factor mobility (e.g., light manufacturing) have less of an incentive to engage broadly in stakeholder dialogue because it is often cheaper and easier to simply exit a difficult production context than to engage with workers or community members.

Where such dialogues do occur, they tend to be firm-specific in nature rather than industry-wide (Locke 2013). By contrast, firms in industries with high "sunk costs" (e.g., extractives) are likely to consult more widely, both because of the non-mobility of resources and because of existing precedents in international law. For example, the International Labour Organization (ILO) Convention 169 requires states parties to regulate corporate involvement in mineral resource extraction so that consultation with local indigenous populations occurs in the course of securing access rights to mining sites (Costanza 2015).

Aware of research by Mark Anner (2011), who has found that labor rights promotion through transnationalized human rights activism is more prevalent the closer the industrial base is to the "market" for conscientious consumption, we had anticipated that stakeholder dialogues would be prevalent in regions such Central America and the Caribbean as well as the northern regions of South America because these are closer to the United States than are regions such as South Asia. But analysis of country-level data in the BHRRC database revealed that reported consultation is not geographically clustered in proximity to export markets in that region or others. Dialogue happens in roughly equal numbers across regions *within sectors* (i.e., in African extractives, 613 cases; in Asian extractives 442; in Latin American extractives 460 vs. in African light manufacturing 41 cases; in Latin American manufacturing 30; and in Asia 95). This is counter to expectations in the political economy literature about "neighborhood" effects and yields a more complex picture of the relationship between media coverage of BHR issues, related NGO activism, and database building and analysis—all of which have their relative limits.

Stepping back from the data, the larger explanation for concentration in extractives is more likely structural in nature. Richard Locke (2013) has argued that a strategy of iterated engagement with industry partners (rather than a punitive compliance model predicated on the threat of exit) leads to enhanced labor rights outcomes, but he notes that the dominant model is still a cut-and-run strategy of enforcing labor rights by "exiting" trouble spots. (This is more common in light manufacturing than in the extractive sector, where the sunk cost of investment is higher and the resource itself being mined or otherwise extracted is geographically restricted to that site.) In light industries where inputs are highly portable, low-wage workers are infinitely substitutable, labor unions are scarce and threatened, and workers may view a job albeit with poor working conditions as preferable to no work at all, the odds are that consultation is unlikely to happen, at least publicly. This leads to the possibility that dialogues involving community members along with some of the most vulnerable workers in global supply chains are either not happening or are happening in ways distinct from the extractive sector–based consultations that dominate the existing literature on stakeholder consultation.

ASSESSING DATA FROM COMPARABLE STUDIES

Several recent publications among the practitioner-generated "gray literature" identify trends that closely mirror those identified in our analysis of BHRRC data. A first study, published in 2015 by OECD Watch (a global network of monitoring and advocacy organizations), analyzes 250 cases of grievances processed over a 15-year period through the Organisation for Economic Cooperation and Development (OECD) National Contact Point reporting system.[12] Its findings parallel our analysis of the BHRRC's database, both in terms of dispersion among industrial sectors (top-heavy in extractives) and in terms of regional trends.

Specifically, the complaints filed from 2001 to 2015 through the OECD National Contact Point system are disproportionately concentrated in the extractive sector (i.e., 57 cases in mining, 33 in oil and gas) versus light manufacturing (i.e., 13 garment companies). Another 52 cases were lodged through the OECD process in sectors not covered in our own analysis of BHRRC data (i.e., there were 52 other cases divided between the financial services sector, which accounted for 29 cases, and the agricultural sector, which accounted for 23 cases). The regional dispersion of complaints was roughly evenly distributed between Latin America (42 cases) and Asia (54 cases). (See OECD Watch 2015, 12, 14, 15.)

A second study, published in 2014 by SHIFT (the nongovernmental research group founded to work with business and governments on follow-up to the Ruggie Principles), analyzes reporting on the corporate disclosure policies of 43 companies covering the period October 2013 through March 2014.[13] In SHIFT's sample, companies in the extractive industry have the strongest level of engagement with "affected stakeholders when assessing impacts" (SHIFT 2014a, 7) but few companies in other sectors tend to disclose information on engagement with stakeholders in the supply chain. Apparel, food, beverage, agriculture, and Internet communications technology sectors (in that order) tend to disclose progressively less information on consultations (SHIFT 2014a, 4). Disclosure of supply chain information in the apparel and food sectors is due in large part to consumer demand for evidence that companies are attempting to improve working conditions in those sectors. Corporate concern with retaining brand value (particularly of high-prestige brands) often underlies the decision to disclose (Deitelhoff and Wolf 2013, 239).

A third study, conducted from 2014 to 2016 and published in 2017 by MSI Integrity and the Duke University Human Rights Center, assessed 45 multi-stakeholder initiative (MSI) standard-setting organizations active in 170 countries on six continents. Together, these organizations oversee voluntary compliance standards integral to the activities of over 9,000 companies in multiple industries (Collins, Evans, Hung, and Katzenstein 2017, 4). The spread of those industries closely parallels what we identified in the BHRRC data—namely, within the MSI Integrity sample, agriculture, forestry and fishing, mining and energy, and consumer goods accounted for 90% of the sectors covered by MSIs. Mining accounted for nearly a third (28%) of all MSIs analyzed in this study and agriculture, forestry, and fishing accounted for another 44%, leaving just 17 percent for "consumer goods," which itself is a category divided among companies in the automotive sector and the personal and consumer goods sectors (Collins et al. 2017, 4; see also MSI Integrity methodology document 2017, 20).[14]

The MSI Integrity study focuses on the structure of the standards mechanisms themselves (i.e., how they are governed) rather than the "rigor, depth, or impact of MSI standards" or their impact on firms or communities (Collins et al. 2017, 8). It does not delve into the impact of sanctioning or the effectiveness of complaint or redress procedures (Collins et al. 2017, 8). But the report does reveal that the particular MSIs analyzed are characterized by a lack of meaningful engagement with "communities affected by the operations of participating companies" (Collins et al. 2017, 3), not only in terms of internal decision-making but also in terms of vehicles for sanctioning member companies or for addressing community complaints or determining the nature of redress.

Similarly, the 2015 OECD Watch study, which was tellingly entitled *Remedy Remains Rare*, finds that the system for receiving and responding to allegations of corporate human rights violations has "historically failed to serve as an effective forum for remedy" (OECD Watch 2015, 9).

Just 1% of the complaints filed have resulted in an outcome that "directly improved conditions for the victims of corporate misconduct" (OECD Watch 2015, 19). The complaint system is unable to "bring an end to corporate misconduct or provide remedy for past or ongoing abuses, leaving complainants in the same or worse position as they were in before they filed their complaint." It also tends not to accept allegations of potential future harms (OECD Watch 2015, 5).

The SHIFT report of 2014 also acknowledges that corporate disclosure on human rights policies and impact remains overly general and untethered from robust community impact analysis: "Many companies disclose information about stakeholder engagement processes. However, they often describe processes that are led by corporate headquarters. Few companies disclose evidence of how they engage affected stakeholders in their processes for assessing impacts or tracking the effectiveness of actions taken" (2014a, 3; see also p. 11). Two-thirds of the companies in the sample analyzed by SHIFT were unclear in articulating how they determined the nature of risk, whether "in relation to rights holders, as against risk to the business alone; or whether [risk] encompasses all internationally-recognized human rights, rather than starting from a sub-set of human rights, such as labor rights" (2014a, 10). Notably, this report includes limited discussion of whether or not local peoples' own perceptions of norms and rights are relevant to interpreting risk.

Two of the key variables identified in Chapter 1 as limiting the scope of remedy available to grassroots participants in stakeholder dialogue (i.e. noninclusive rules-setting and foreshortened timeframe) are central to the problems discussed in the reports by OECD Watch, SHIFT, and MSI Integrity. The rules of such consultation are disadvantageous to people in local communities, who are often poor and have limited access to the formal legal system (see also Skinner, McCorquodale, De Schutter, and Lambe 2013). Taking part in consultations is inordinately expensive and impractical for people at the grassroots level.

For example, if members of a local community in a developing country raise a complaint against a European company through the OECD National Contact Point system, the national contact point in the company's *home* country may require the local community members in the *host* country to fund their own travel to a consultation that will take place in the home jurisdiction (in Europe) and to fund translation costs while there. The threshold for evidence submitted in non-judicial grievance procedures, the criteria used to assess the merits of the case, and the requirement that complainants exhaust all other legal avenues combine to make it difficult if not impossible for local community members to use the OECD complaint system. Moreover, people in local communities are frequently blocked from equal access to evidence or other information integral to the complaint process. Even when mediated agreements are brokered, they are poorly monitored through the OECD process.

When community members lack either an understanding or ability to influence the terms of consultation, the scope and staging of remedy can lead to unevenness in the extent to which they experience any potential benefits from the process (Fulton, Ha, Karimian, Lerner, Meier, and Plessis 2015, 1–2, 24–29). Limits on the accessibility, impartiality, transparency, and predictability of consultation can essentially hollow out participation and stymie the movement "upstream" of normative understandings of human rights and corresponding remedies that reflect community interests and concerns. The worst-case scenario is that participation becomes ritualized at the expense of meaningful outcomes for communities. Networks of NGOs and community organizations either involved in dialogues themselves or observing them already report such trends (de Felice 2014). Hence, the qualitative case studies central to this book explicitly seek to uncover grassroots-level people's perceptions of the context for stakeholder consultation and their perspectives on different strategies for ensuring economic rights protection and remedies for people in communities affected by business activity. A brief discussion of the logic of case study design and the nature of related data gathering follows.

SETTING THE STAGE FOR INQUIRY: CASE SELECTION

The comparative case study chapters that follow are based on a "method of difference" research design aimed at exploring social attitudes on stakeholder dialogue in two manufacturing towns in the Dominican Republic. Specifically, I analyze whether or not community members' receptivity to stakeholder dialogue and their broader conceptualizations of remedy for economic rights violations are affected by distinct corporate approaches to BHR practice in the two towns. As we have already seen, the bulk of stakeholder dialogue taking place globally today occurs in the extractive sector, not in light manufacturing, so case studies of communities affected by light manufacturing help fill a key gap in theory and practice on BHR.

Two distinct business models and two distinct approaches to social responsibility on the part of the companies have unfolded in towns less than an hour apart in the Dominican Republic. In Villa Altagracia, a town of roughly 84,000 inhabitants[15] located just over 30 kilometers from the capital, Santo Domingo, the main textile factory—Alta Gracia Apparel—pays triple the national minimum wage to its roughly 200-person labor force and engages in extensive international marketing and branding of the collegiate apparel it produces, based on the living wage or *salario digno* (Adler-Milstein and Kline 2017, 8) earned by its workers. This business model is the result of years of worker-driven social responsibility (WSR) efforts on the part of Alta Gracia workers and their allies internationally. Multi-year field studies carried out by a research team at Georgetown University have highlighted gains in worker well-being (Kline 2010; Kline and Soule 2011, 2014), which they argue have a multiplier effect throughout the local community (Kline and Soule 2014, ii; Adler-Milstein and Kline 2017, 28).

In Bonao, a town not quite twice the size of Villa Altagracia (with over 125,000 inhabitants) roughly 37 kilometers farther north of Santo Domingo, a major spinning mill owned by the Hanes Corporation employs upwards of 3,000 workers (Hanes Corporation 2017) to produce fabric in four shifts that operate around the clock. The emphasis in this factory is on lean production (i.e., minimizing waste and maximizing quality), and base pay is the conventional Dominican minimum wage. Hanes covers the cost of ongoing worker education coordinated by the

Ministry of Labor's National Technical and Professional Training Institute (Instituto Nacional de Formación Técnico Profesional [INFOTEP]). The company also engages in local philanthropy by organizing the biannual visit of a team of American medical volunteers who collaborate with local hospital staff to provide ear, nose, and throat surgery for children in the local community of Bonao. Hanes's model in Bonao involves conventional supply chain monitoring in the BHR mode, with some novel adaptation of training and philanthropic activities.

To date, there has been no academic attempt to explore *community perceptions* of corporate responsibility and social well-being in comparative perspective between these two towns. How do people evaluate their own well-being in the face of two distinct approaches to doing business in their respective communities (i.e., WSR vs. conventional BHR)? Is there a substantive difference between perceptions of corporate responsibility in a living wage community and a non–living wage community? Is there variation in the degree to which non-workers are receptive to dialogue with companies? Are there significantly different perceptions of remedy? Why or why not? These were the central questions motivating my conversations with community members in each town.

By holding country, economic sector, and export market (collegiate apparel) constant, the case study design enabled me to focus on whether or not differences in corporate practices in the two communities produce variation in local peoples' attitudes regarding stakeholder dialogue processes and remedy. One of the cases (Villa Altagracia) selected for comparison is a crucial case (VanEvera 1997, 66-67), centered on a town where the leading industrial manufacturer is internationally renowned for paying triple the national minimum wage. Collegiate apparel itself is a market segment in which consumer awareness of labor rights issues tends to be somewhat higher on average than in other segments of the population (Elkins and Hertel 2011). If we could anticipate a context ripe for broader community engagement within the light manufacturing sector or for corporate action on remedy for economic rights violations, this would intuitively be it (Adler-Milstein and Kline 2017). The other case selected for comparison (Bonao) is a town in the same sub-region of the same country with a leading firm in the same industrial sector (textiles) that exports to the same market niche (collegiate apparel) but pays a standard wage while engaging in community philanthropy. The two cases thus offer a vehicle for exploring whether variation in approaches to protecting human rights by the two similarly situated firms affects the receptivity of people in the respective communities to potential stakeholder engagement along with their views on economic rights and remedy.

The interviews[16] central to this set of comparative case studies aim at disentangling the factors that affect an individual person's sense of well-being and corresponding differences in notions of remedy within communities impacted by light manufacturing. A respondent's subjective social status (SSS), defined as an "individual's subjective evaluation of their overall social status" (Landefeld, Burmaster, Rehkopf, Syme, Lahiff, Adler-Milstein, and Fernald 2014, 92), serves as a proxy for how individual people evaluate their well-being in the face of corporate activity within a given community. My research builds on earlier public health–related research in the "triple wage" community (see Landefeld et al. 2014) that analyzed how changes in wages affect SSS and how SSS, in turn, affects health outcomes through a controlled comparison with factory workers in a non–triple wage factory in another town. Existing research (Kline 2010; Kline and Soule 2011, 2014; Landefeld et al. 2014) has centered on workers' perceptions of their own well-being, either assuming that the benefits of income gains would trickle down to other community members (Kline and Soule 2014, ii) or excluding community members from analysis (Landefeld et al. 2014).

The tripling of the minimum wage salary at the factory in the first town provided the context for a natural experiment that would enable researchers to explore the causal relationship among the wages, SSS, and health among workers in both settings. I make a two-step move beyond previous scholarship on the crucial case of the Alta Gracia Corporation: (1) by moving empirically *beyond workers* to include community members more broadly in my analysis of the sources of change in SSS and (2) by moving beyond health outcomes to explore a wider range of economic rights issues and related notions of remedy within this broader population. The result is a bottom-up view of the complex community contexts in which the forging of remedy plays out. The case study chapters that follow take stock of what is happening on the ground in the broader community in each town as these two distinct approaches to manufacturing unfold. The lessons learned about the challenge of defining the terms of community engagement and effective remedy in the context of light manufacturing are central to Chapter 6 (i.e., the policy chapter that follows the two cases), which draws on qualitative data also collected in 2017 during an international conference on stakeholder engagement.[17]

DATA GATHERING AT THE FIELD LEVEL

To reveal the complex mix of factors that influence personal receptivity to consultation and perceptions of the prospects for remedy, the case study chapters that follow delve into original qualitative data from 43 interviews and participant observation conducted in the Dominican Republic over a three-week-long field study in 2017. From late May through mid-June 2017, I interviewed residents of Villa Altagracia and Bonao while also observing daily life in both communities. In Santo Domingo, I also interviewed a national union leader familiar with labor rights advocates in both communities. Using snowball sampling, I expanded my contact base from an initial set of key NGO, union, and advocacy contacts to whom I had reached out months in advance. The friends, neighbors, and colleagues of these people, in turn, introduced me to others, whom I sought to engage in systematic interviews.

My aim when conducting formal interviews was not to create a randomized sample from which to generalize about trends within the national population, but instead to explore the complexity of the comparative community contexts along with differences in SSS and perceptions of responsibility for remedy among people in each community.[18] A set of interview questions (see Appendix 1 for English version and Appendix 2 for Spanish version) created a framework for engaging people in structured conversation in which they shared their life and work histories; their attitudes regarding their own employment opportunities and social development prospects; their opinions on the nature of responsibility for community well-being; their knowledge of corporate activity in their community; and their experiences (if any) of direct involvement either as a worker or, more often, as an observer of companies producing textiles for export in their community. I interviewed adults ranging from elderly to college-aged. Some were retired; others worked in various types of businesses (formal and informal). Others cared for family members; others were actively seeking work. The typical interview lasted 30 minutes, and my questions (some open-ended, some multiple choice) helped guide the conversation.

While in Santo Domingo, I also sought quantitative data and published reports on the mandates of three government ministries with activities in Villa Altagracia and Bonao from staff of those agencies: INFOTEP; the Ministerio de Economía, Planificación y Desarrollo (MEPyD; Ministry of Economic Planning and Development); and the Ministerio de la Mujer (Ministry of Women). I used these data to develop a fuller picture of the policy context in each town.

PLACING THE LOCAL CONTEXT IN PERSPECTIVE

In Villa Altagracia and Bonao, people face multiple challenges in their everyday lives. They determine their short- and long-term priorities influenced by these constraints. The positions they take in negotiations of any kind—whether with friends, neighbors, or representatives of business or government—are influenced not only by these constraints but also by their perceptions of the rights they have. Conventional stakeholder dialogue practices, however, lack robust mechanisms for engaging community members in setting the scope and terms of their interaction with companies, in part because of a dearth of knowledge about or willingness to address the structural roots of poverty pervasive in communities along the global supply chain.

Whether sitting on a porch, at a food stall alongside the highway, or in a mechanic's shop, my interviews with people in Villa Altagracia and Bonao opened by inviting participants to share their life history, framed around their own employment history. Respondents varied widely in their levels of education, from functionally illiterate to university educated. Some of the people I interviewed would preemptively caution that they were not sure they would answer any of my questions "correctly," to which I would reply that there was no way they could answer incorrectly if they were sharing their own story.

My ability to gauge the SSS of anyone I interviewed thus hinged on creating an interview setting in which the person could place her or his life story in context vis-à-vis other people in the community and against the background of larger trends that shape the potential for personal socioeconomic well-being or distress in light manufacturing towns like Villa Altagracia and Bonao. Only then could we segue to broader questions centered on their perceptions of community well-being or employment opportunities and social welfare support. As the interview progressed, we could then explore more complex themes, including their knowledge of corporate activities in the area; awareness of stakeholder dialogue practices; and opinions on the nature and scope of responsibility for community well-being borne by government, the private sector, unions, or people in civil society themselves.

Engaging in conversation, observation, and participation in everyday life in each setting was integral to the process of understanding the local context. In order not to bias my sample, I explicitly avoided visiting the factory zone in either town, nor did I interview company management in either town. Most of the people I interviewed in Villa Altagracia had connections to the factory through friends or family (i.e., 15 of the 27 people I interviewed in Villa Altagracia were family members—either nuclear or extended—of factory employees; two worked directly for the company; another 10 had no connection to the factory). By contrast, in Bonao only one of the people I interviewed worked for Hanes; two had family members who either currently worked for or had previously worked for the company; another two people had children who had received free operations through the company's philanthropy; and the remaining 11 had no direct connection to the company.

The paired case studies central to this book reveal the complexity of the local contexts in which stakeholder dialogue unfolds. As Richard Locke (2013, 80) has observed, in light manufacturing settings globally "most capability-building initiatives embrace a highly-technocratic approach to supply chain factory 'upgrading' that may help address certain problems . . . but eschews both more *distributive issues* and concerns with *enabling rights* that are core to labor standards enforcement" (emphasis added). The Alta Gracia factory model is unique because it explicitly seeks to distribute the gains from production more fully to workers by paying a living wage. Through a model of WSR that shapes factory governance, Alta Gracia challenges the low-wage textile industry model directly (Adler-Milstein and Kline 2017)—hence my interest in analyzing community perspectives on remedy in a town where at least one employer is explicitly committed to distributive justice issues.

The case study chapters that follow thus put people in the community of Villa Altagracia *beyond the factory* into comparative perspective with people living in the neighborhood of Las Delicias in Bonao, whose lives are impacted directly or indirectly by the Hanes factory. In the process, I foreground the "distributive" and "enabling rights" that Locke argues have been ignored in the literature. Distributive rights are integral to people's own sense of their SSS. More broad-based and nuanced analysis of SSS, in turn, helps shed light on people's perceptions of which entities (states, corporations, groups in civil society, or individuals themselves) are responsible for safeguarding community well-being.

As these paired case study chapters reveal, my conversations with a broader range of people than workers alone help shed light on a complex picture of needs, rights, and remedies beyond the immediate production context. These conversations have also revealed considerable variation in the willingness of average people to engage with the companies in their communities in the interest of protecting and promoting economic rights.

CHAPTER 4

The People Beyond the Tag

Stakeholder Perceptions in Villa Altagracia

Politicians always say if they win, they'll make sure that this or that factory stays open, and they never do anything.
—Former factory worker, now food vendor[1]

No one worries about anyone else . . . for the majority of people, money is concentrated among two or three people; for poor people like us, we don't matter at all to these other people.
—Grandmother and homemaker[2]

We're positioning ourselves for when development happens. . . . With the theme of globalization, we understand how the world's development [happens] and we want a life like that. So there are a lot of people who want another type of life and although we don't have it, at some point, it'll develop and we'll achieve it.
—Independent artist[3]

WALK THROUGH ANY AMERICAN RETAIL outlet and the faces displayed in advertising are typically the glamorous ones of models paid to wear the clothes you'll buy, not the faces of the people who sewed the clothing. The exception is Alta Gracia Apparel, marketed centrally to college students and displayed alongside advertising photographs of workers themselves, whose faces also appear on the hangtag affixed to the garments they sew. These people live and work in a town that bears the same name as the brand of clothing they make: Villa Altagracia, just over 30 kilometers from Santo Domingo in the Dominican Republic.

Collegiate apparel sales are integral to the business model of Alta Gracia Apparel; the company itself is a unique outcome of the Workers' Rights Consortium (WRC)'s consistent advocacy of a living wage for garment workers and of a worker-driven social responsibility model that the factory embodies. Locally based workers' rights organizations have played a crucial albeit often ignored role in raising workers' consciousness about basic labor rights internationally (Ackerly 2018). Adler-Milstein and Kline's 2017 book tells the story of the key role workers in this factory have played in helping build an alternative business model centered on a living wage that is triple that of the standard Dominican wage. At the time the factory opened in 2010, Alta Gracia workers earned the equivalent in Dominican pesos of $2.83 US dollars per hour versus the legal Dominican minimum wage of 83 cents per hour (Adler-Milstein and Kline 2017, 9, based on 2010 figures). The monthly full-time wage of an Alta Gracia worker was $497.34 (in 2010 dollars) versus $147.95 per month for standard minimum wage worker, or a difference in annual income of $4,662 (net) for the Alta Gracia worker versus a conventional minimum wage annual salary of $1,474 (Workers' Rights Consortium 2010, 6; Workers' Rights Consortium 2013, 15).[4]

Worker speaking tours on college and university campuses throughout the United States (including 19 in six major cities in 2011 alone, facilitated by the WRC and United Students Against Sweatshops) along with visits from student tours to workers' homes over the past six years have helped foster brand awareness among consumers in industrialized countries (Adler-Milstein and Kline 2017, 78–90). The WRC has an office in the Alta Gracia factory itself and has collaborated directly with local Dominican labor unions to actively monitor workers' rights on an ongoing basis (Adler-Milstein and Kline 2017, 23–26, 100–105, 173).

But advocacy efforts around collegiate apparel have tended to focus principally on changing conditions *within* factory spaces and in the lives of workers employed directly, and less on broader, community-level economic rights fulfillment. This begs the question of whether or not members of the community *beyond* the factory perceive themselves as stakeholders.[5] The role community members play in defining the terms of engagement with corporations and the scope of remedy remains unclear theoretically and in practice.

Workers in the Alta Gracia factory are by no means the majority of people in town. There are barely 200 of them in a population of roughly 84,000 people (Adler-Milstein and Kline 2017, 8; República Dominicana Sistema Interactiva de Consulta Censo [SICEN] 2010), but they "are perceived to be among the most well-off residents in town" according to Adler-Milstein and Kline (2017, 122). The Dominican peso (RD peso) has weakened significantly in the years since the 2010 WRC living wage calculation was made central to collective bargaining by Alta Gracia workers. The peso to US dollar exchange range went from 36.7 RD pesos to one US dollar in 2010 to 47.6 pesos to the dollar today, and rising inflation has cut into the purchasing power not only of Alta Gracia workers but of others in the community as well.[6]

Do residents of Villa Altagracia perceive themselves as having a stake in the fate of the Alta Gracia factory or a role for themselves in dialogue with business more generally? Do they view themselves as entitled to remedy from harm or to benefits generated by any of the factories in their community? Why or why not? This chapter moves beyond observations of wealth transfer among a relatively small number of people in the community to explore whether variation in subjective socioeconomic status (SSS) affects a broader range of peoples' perceptions of their own rights; their relationships to companies active in their community (including Alta Gracia); and their notions of who is responsible for remedy when rights are abused or not fulfilled adequately.

Based on conversations with people in Villa Altagracia in May and June 2017,[7] this chapter explores relatively uniform concern with unemployment and underemployment but considerable variation in knowledge of corporate activities in the area. There is also variation in awareness of stakeholder dialogue practices and opinions about the nature and scope of responsibility for community well-being—including whether it should be borne by government, the private sector, unions, or people in society themselves. This variation, even in a context that would seem ideal for a new form of stakeholder engagement, reveals the challenge of determining who is involved in (or outside of) such consultations and why and how stakeholder engagement does (or does not) contribute to changing the reality of life on the ground for people in manufacturing communities.

GLOBAL PRESSURES, LOCAL REALITIES: INDUSTRIAL DECLINE AND EVERYDAY LIFE

Like many other places that have endured steep increases in unemployment in the wake of industrial downsizing, the overall economy of Villa Altagracia is depressed and the majority of those I interviewed characterized living standards as worsening and employment opportunities as scarce. Because several major manufacturing companies have left the main export processing manufacturing zone in Villa Altagracia since the early 2000s—including a large textile manufacturing plant that dismissed over 3,500 workers when it closed in 2008, along with a sugar refinery and a paper factory—the quality of work in the model Alta Gracia Apparel factory is dwarfed by the enormity of the unemployment problem beyond the factory's walls and by attendant social pressures.

The Dominican Republic was the first country in the Caribbean to transition to democracy (in 1978) and has experienced relative political stability over time since then (Brea, Duarte, and Seligson 2005, 54) but the contemporary political context is marked by extensive racial fragmentation and party patronage (Fine and Petrozziello 2017, 2, 85). While ranked as an upper-middle-income developing country according to World Bank data (2011, cited in Landefeld, Burmaster, Rehkopf, Syme, Lahiff, Adler-Milstein, and Fernald 2014, 93), the Dominican Republic is an increasingly unequal one. The 2015 UN Development Programme's *Human Development Report* revealed that gap starkly: the country's gross national income per capita increased by 143.8% between 1980 and 2014 (United Nations Development Programme 2015, 2) yet the real poverty rate increased from 47% to 70% from 1984 to 1991 (Fine and Petrozziello 2017, 73). Moreover, the "incidence and burden of chronic diseases is rapidly increasing" and unevenly distributed, concentrated among the poor (Landefeld et al. 2014, 95). The country's 2015 Human Development Index ranking of 0.715 was only slightly below the regional average of 0.748 (United Nations Development Programme 2015, 4), yet when adjusted for inequality it was 24% lower, at 0.546 (United Nations Development Programme 2015, 5) during the same year.

Like many other countries in the developing world, the Dominican Republic shifted its development strategy in the wake of structural adjustment in the early 1980s away from agricultural production (primarily sugar) toward export-oriented manufacturing and service-sector development. The Dominican textile sector has been integral to this shift, but growth in the sector has been predicated largely on low wages and preferential tax policies for investment (Fine and Petrozziello 2017). In the decade from 2001 to 2011, garment workers in the Dominican Republic experienced a 23.74% decrease in wages (Workers' Rights Consortium 2013, 3) and the country's overall position in the industry declined, slipping from the fifth largest source of apparel for the US market in 2000 to 21st by 2011, thus losing 80% of its market share over the same decade (Workers' Rights Consortium 2013, 10).

A smaller percentage of the Dominican labor force is unionized than in other countries in the region: historically, unions have never represented more than 15% of the labor force (Fine and Petrozziello 2017, 72), and in 2006, only 6.3% of men and 2.9% of women were members of a union (Brea et al. 2006, 158). Gender inequality in the Dominican Republic is particularly pronounced: while Dominican women on average live longer and attend school longer than men, they nevertheless reap slightly more than half the gross national income per capita of Dominican men (i.e., $8,860 per year vs. $14,903 for men) and the adolescent birthrate is 99.6 per every 1,000 women aged 15 to 19, in contrast to the regional average of 68.3 per 1,000 (United Nations Development Programme 2015, 6).

Dominican workers have tended not to enjoy the full protection of labor laws that exist *de jure* but are often violated *de facto* routinely. Schrank (2013, 314) argues—and my interview data corroborate—that labor organizing in the textile sector remains a challenge in part because of hostility toward organized labor on the part of some free-trade-zone workers[8] and community members; in part because of variation in the quality of work in Dominican free-trade zones (Schrank 2011, 427); and in part because of variation in individual people's perceptions of the responsibility unions bear for more general social well-being.

In the wake of factory closings in Villa Altagracia over the past decade and the corresponding lack of work for the thousands of people who lost their jobs, many of those formerly employed in the city's industrial sector have gravitated toward service work in the capital, Santo Domingo, an hour away by bus. Some work as domestic helpers, others as day laborers on construction sites. Nearly all of the respondents

I spoke with described Villa Altagracia as "a town without work, where people have to leave to find work."[9] As a local bus driver explained:

> The well-being of people dropped hugely in about 2000 because there wasn't work. In order to work, people have to go to Santo Domingo. They do domestic work—women in houses—there are men who work in construction of businesses.... People leave on Mondays and stay until Fridays because they stay in homes or construction sites ... [At the construction sites] no one pays for this lodging, they just stay there.... They don't have electricity, they don't have beds, they don't have bathrooms.... It's really precarious.[10]

A young college student echoed this sentiment: "People have to migrate for work opportunities and if one doesn't have the opportunity to study ... they have to pick up a pole or a machete and break up the earth to make a patio or something—and that makes their life a lot less comfortable. If they have kids, they have to make do with what they have."[11]

Commuting costs are high relative to the low wages in the service sector, so many of the people employed in service work resort to living at their employer's homes or on construction sites, leaving their children in the care of friends or relatives back in Villa Altagracia for most of the workweek. A former worker at the BJ&B cap factory, which closed in 2000, was among those I interviewed, one of the thousands who lost her job. Initially, she too, worked as a domestic servant but now has a small food stand where she sells pastries (she ground yucca sitting on her porch as we spoke) and cares for her grandchildren so that her daughter can commute daily for a second-shift job (i.e., 3 to 10 p.m.) in Santo Domingo, returning close to midnight. The children spend the night with the grandmother; she brings them back to their mother's home in the morning and gets them ready for school before resuming her food preparation.

Of the three people I interviewed who were making food or handicraft products as we talked, only this woman had received a small business loan through the government, but neither she nor other small business owners were involved in formal microenterprise promotion programs that would have afforded them ongoing access to credit, training, or formal marketing opportunities. As another former BJ&B worker-turned-food-vendor explained:

> I worked in BJ&B until they folded. Then I worked in a free-trade zone in the capital for about a year. But I left because it was difficult—the commute, the danger [of crime for those who commute]—so I figured it was better here . . . I look for work, but still can't find it. So, I've put together this little cafeteria and I have a little billiard table and I serve people. This is my friend [gesturing to another woman preparing food in the back of the food stall]. She lives here and has four children. She said she doesn't have work also; so I said, "Come on, we'll both eat," so we work together.[12]

Both are single mothers.

All three of the former BJ&B workers I interviewed commented on the contraction of industry and the ripple effects on the wider economy since the early 2000s. The oldest of them (the grandmother who cared for her grandchildren and sold yucca pastries) observed:

> After the factory zone shut down, there hasn't been a lot of work. But before there was more work, because there were people who made and brought food for the workers, people who worked doing ironing, others who were drivers for them—and now there's a lot of unemployment . . . They said that they were going to reopen the free-trade zone, but the president would have to come here, and bring investors. They had everything ready and 4,000 people had brought their identity cards to apply for work, but I don't know what happened. . . . There's been nothing concrete.[13]

A second former BJ&B worker (a grandmother who cared for the child of a daughter who worked in an administrative job in Santo Domingo) sold used clothing but noted that

> . . . there is no source of work other than the free-trade zone or the citrus production plant. . . . People can work in shops, and lottery stalls, in the capital. . . . But there's not much. If the government would lend a hand . . . and bring businesses, then there'd be a source of employment but on the contrary, no. . . . Look at the majority of people: What can they control? When you look in their faces, they're beaten down.[14]

Echoing the bare-bones nature of small businesses in Villa Altagracia, the owner of a small tire repair workshop explained, "Here there are a lot of businesses; you don't even realize it, but the majority of them are broke."[15]

Even for people who commuted to Santo Domingo for work, high commuting costs often outweighed what they earned. One young mother I interviewed opted to serve as a housemaid instead of working in the garment industry herself. As she recounted, she had "worked in a small workshop in Santo Domingo making ballet costumes. I'd work three months and then I'd stop and then they'd rehire me—it was temporary work." She opted to do housekeeping instead "because the owner of the shop wouldn't cover the cost of my commute and I didn't earn enough to cover the cost of transport, so I had to leave the job. Every day I paid 170 pesos [$3.50 US dollars] for the commute and I earned 3,000 pesos [$62.50 US] monthly"—which meant that the cost of her commute for a typical 20-day month of work would average 3,400 pesos ($71 US). "Initially the owner paid half," she recalled, "but then he wouldn't. . . . Here, I'm earning more than I earned in the workshop, and I come on foot."[16]

Prohibitive commuting costs leave even people with higher skills training marooned in Villa Altagracia. As the husband of an Alta Gracia worker explained, underemployment is pervasive. Trained in industrial and home refrigeration maintenance, he cannot find work in Villa Altagracia:

> . . . the work situation here is really difficult—there are only jobs in the companies in the free-trade zone, and there are almost no other firms. Shops and other things like this pay almost nothing. . . . . I would have to travel to Santo Domingo, but the problem is, they pay very little, too—7,000 pesos [$146/month] per month, if that. The commute cost is 4,000 pesos per month [$83/month], so that doesn't leave you anything. . . . There are many unemployed people here, every day, you see the news that people here are without work, so they take a car or motorcycle and make it a taxi, to get by.[17]

In this particular case, the man had his wife's steady salary as a source of household income, in contrast to others I interviewed.

EVIDENCE OF EVERYDAY CHALLENGES

Dominicans rely on public assistance in the form of conditional cash transfers at a much higher level than any other country in the region (Brea et al. 2006, xviiii), and in many of my interviews in Villa Altagracia, people referenced the specific types of public assistance they receive, from subsidized electricity to food vouchers. "They do a census so they can apportion the cards for food, for cooking, for electricity. . . . I get this help," a former BJ&B worker explained, "and also a benefit of 300 pesos [$6.30 US] a month for my youngest daughter, who is in high school. Some benefits come monthly and others on alternating months."[18] Others rely on relatives who live abroad to support them through remittances. I conducted one interview on the front porch of a house undergoing second-story renovations, paid for by sisters who lived and worked in Chile and Argentina.

But many other homes are in poor condition. Describing the humble homes that Alta Gracia workers formerly occupied, Adler-Milstein and Kline point out that "living conditions like these are not just a baseline for Alta Gracia workers, but the current reality for the rest of the community" (2017, 84). Paying a living wage to Alta Gracia workers generates "economic multiplier effects" that the authors argue reach "other households in the local community" (2017, 118), so "[they] contribute to the economic health of the larger community—sparking the foundation and growth of new small businesses and putting more money into the pockets of existing community businesses" (2017, 191). Alta Gracia workers are thus "able to repay the loans and assistance they obtained from relatives during unemployed periods and then offer similar assistance to others" (2017, 122). The company's workers have negotiated collective bargaining agreements with management since the factory opened in 2010, and Adler-Milstein and Kline note that workers have included provisions for community development such as "a school supply fund for other local community children, payment for a community service day, and a commitment to help open a learning center to improve access to technology and the internet" (2017, 104). The authors recount one worker's observation that "she feels an obligation to those outside of Alta Gracia—those who have no work or means of support, especially mothers who must leave town to seek work to feed their families. Without that interview ticket [which enabled her to secure her own position at the factory], she likely would've joined them" (2017, 163).

Indeed, for people not employed by Alta Gracia, the phrase *"no alcance"* (it doesn't go far enough) was a common refrain in my interviews—whether in relation to government welfare or to pay for work. This and growing perceptions of corruption in major state institutions (courts, police) and ever-increasing levels of public concern over worsening personal safety were common themes across ages and genders. Villa Altagracia is the most violent city in the Dominican Republic (US Department of State 2014), and this has resulted in high levels of isolation among residents, who often commented on the weakness or non-responsiveness of local government officials. These observations at the micro level are mirrored in national public opinion polls, which reveal the steady erosion in public opinion of state legitimacy since roughly 2010 (Morgan, Espinal, and Seligson 2006).

Personal concern with violence was a repeated reference in my conversations in Villa Altagracia, where robbery and assault were an everyday factor of life. As a retired garment worker explained:

> ... there are so many people who want to work yet there isn't work now for them, and for this reason I think there is now so much more delinquency because these people can't work. So they look instead for a way to make an easy buck—and they stand in doorways and rob people . . . I guard myself when someone I don't know puts out their hand; I have fear out there. I go to church, I read the Bible . . . but there are people who don't think like this, and they take what they can in order to survive.[19]

Corruption was another frequent touchstone, particularly given that my fieldwork coincided with a series of high-profile arrests of public officials and business leaders involved in widespread corruption scandals related to government contracting for infrastructure projects (Pineda 2017; Lopez 2017). Vignettes of corruption at the local level were often the most galling for those I spoke with in Villa Altagracia. The poverty line used to determine who receives government subsidies for food, cooking gas, electricity, and other necessities "isn't calculated seriously," an Alta Gracia worker explained: "If you get ten people together, and seven of them earn less than 5,000 pesos a month [roughly $105 per month], among those seven there will be five of them who don't have the solidarity card," the identification card necessary to access subsidized benefits. "It's political," she sighed. "It's through friends. If I'm a friend of *Fulano de Tal* [John Doe], then I find five people in my community who are my friends or family members, and these other people all get the solidarity cards, too." The cards are supposed to be allocated through social workers, "but they're 'social workers' in quotes. These people are employed by the mayor's office . . . It happens all over the country. It's not exceptional here. In this, we're all equal; we're not better or worse. In the whole country, you're going to see the same thing."[20]

One of the most obvious indicators of weak state capacity are frequent power outages. In Villa Altagracia, the town's entire power grid routinely shuts down from 5 to 11 p.m. daily. Residents thus plan their activities around the anticipated power outages. For families without access to a generator, the best way to spend these unlit evening hours is often on the street, doing shopping or errands in stores that have generators (larger ones, such as grocery stores, often do; smaller ones, such as a clothing boutique where I interviewed one woman in a central business district of Villa Altagracia, do not). Younger people frequently spend the later evening hours visiting with friends on the streets. Music blares and motorcycles or scooters cruise (loudly) up and down the main thoroughfares—always just until the power comes back on, when people reenter their homes, watch television, or sit in front of a fan. Then, the streets become almost silent.

Many of those I interviewed in Villa Altagracia remarked on the problems created by lack of electricity, the pervasiveness of crime and vagrancy among young people, and the overall lack of employment opportunities. In a city with a significant crime problem, moving about on unlighted streets at night was a necessity but posed risks. The loss of power in Villa Altagracia is not unique to that city; in Bonao, scheduled power outages occurred either midday or in the early afternoons on a rotating schedule. In Santo Domingo, even in the upscale Gazcue neighborhood, where government offices and tourist hotels are located, the power routinely fails on the weekends as the main national power grid undergoes maintenance. Stretches of up to 12 hours without power occurred throughout my stay in Santo Domingo on several weekends, and the smell of diesel was pronounced in a neighborhood where apartment buildings, hostels, government offices, and hotels all had generators whirring to keep the air conditioning running and modems for the Internet powered up.

Villa Altagracia, however, has also experienced periodic loss of municipal water unparalleled in the capital or in Bonao, where I also conducted interviews (see Chapter 5). Throughout the Dominican Republic (as in many other developing countries), people buy water for drinking or cooking but typically use municipal water for bathing. In some places, however, public piping from reservoirs or other main sources may wash out in heavy rains, thus cutting neighborhoods off from the main water source (apparently a recurrent problem, since the family I stayed with in Villa Altagracia kept large buckets of covered water in reserve for washing in emergencies).

In addition, a lack of municipal services for garbage removal results in pervasive litter (in public spaces, vacant lots, and gutters, and even on the edges of the yards around private homes). This presents a public health hazard on several levels, given the country's ongoing battle with the Zika virus, since mosquitos (the carrier of Zika) breed in stagnant water and trash is a prime source of breeding space. Despite its pervasiveness, litter was not an explicit source of concern in any but one of my interviews with people in Villa Altagracia. Unless I asked about it, trash typically did not come up in conversation—even though it was all around us. In those instances where it was discussed, respondents typically characterized litter as more evidence of government incompetence, corruption, or lack of concern for citizens.

GRASSROOTS PERCEPTIONS OF RESPONSIBILITY FOR COMMUNITY WELL-BEING

Myriad problems of basic infrastructure, then, coupled with the seeming inability or unwillingness of state agents (police, public bureaucrats, politicians) to address unemployment, crime, or other social stresses led to a dominant sense of cynicism and frustration with government on the part of many of those I interviewed. When asked who is responsible for community well-being and given non-exclusive, multiple choice response options—of "people themselves, companies, our government, NGOs [nongovernmental organizations] or churches, someone/something else"—the majority of respondents in Villa Altagracia (15 out of 27 people) answered that while the government should be responsible, it was far from able or willing to fulfill its charge.

Many thought that it was government that would ultimately be able to bring business to their community. As a truck driver observed, "they're the ones who can say, 'We're going to have this type of industry, or make something, or bring people here to set up a factory'—but there's none of this."[21] There was widespread cynicism that government officials would promise factory jobs, purporting to speak on behalf of companies interested in bringing work to the area, but as one housewife noted, "during campaign season, I don't take seriously what they're saying. I don't take part in these rumors."[22]

The fatalism of some was pronounced, as with one woman who noted:

> Everything ultimately is in God's hands, but it's like the president is sleeping—you turn on the television and he's not paying attention to this or that. Politicians are responsible for their province or sector. They should go check on it, and make sure that things move forward.... At least they could come to Villa, listen to complaints, sit down with neighborhood associations, this type of thing—but we have to sit here, quietly.[23]

The owner of a small hardware and feed store pointed out that when someone takes a leadership position in government, "for example, a candidate who wins or a deputy or Council member—there are those he favors, and others not. So it continues the same: for some people it goes well, for others it goes poorly. It isn't the efforts of one himself—it's the political situation that determines well-being."[24]

Others were less fatalistic than indignant. One man, who had lived in the United States for over a decade previously, sharply contrasted the government's role in the two countries both in terms of state capacity and political will to address community well-being: "The government, principally, and their allies are the ones responsible" for well-being in this town, along with "the people that they have under their control, who take everything for themselves and don't share it with anyone. Everything continues the same here. There in the United States, no. There, you work and at the end of the year, you receive your tax refund check. Here, there's nothing like that."[25] The sense that corrupt officials either ignore problems or pocket resources was pervasive. Sitting on her porch alongside a river bank during our interview, a former BJ&B plant worker observed: "Look, this bridge fell down. A government representative came and said: 'You're the responsible one—why don't you fix it?' The government should. It's bad construction! A month ago, it fell down . . . He came and saw it, they talked a lot, but he has not come back. The press has been here. But this is the situation in this town." She sighed, gesturing to the pile of broken concrete slabs and rebar several hundred feet from her front porch.[26]

The owner of a small tire repair business argued that

> . . . a lot of people know me, and the *políticos* will come here and say, "Here, we'll help you with rubber or whatever," or they say, "Get your résumé together so that I can help you." What kind of résumé am I going to get together? You tell me! What kind of résumé is a day laborer going to have? Or a barber, or a tire repair guy? What kind of résumé is this type of worker going to have? *Políticos* ask for this type of thing so they don't have to give you work, they don't have to do anything . . . so the best thing to do is keep your distance, and try to maintain yourself, do what you can for yourself in your own way. People rob us and rob us, and they [people in government] don't do anything.[27]

There was more variation in the extent to which other entities were cited as responsible for community well-being: no one I interviewed in Villa Altagracia explicitly cited either companies or NGOs as solely responsible for it. Just under a quarter of those interviewed in Villa Altagracia replied that individual people themselves should be responsible for community well-being (i.e., six respondents out of 27) in part because they despaired of anyone else helping at all, as the cynicism regarding government mentioned earlier in the chapter indicates. Others, including the college-aged son of an Alta Gracia worker, voiced support for traditional family values as the source of well-being: "things emerge principally through the family—family grounds everything. If a family decides—if a father and mother raise their children well—everything follows from that."[28]

Still others thought that members of society collectively should be responsible for its well-being, by working together (i.e., five respondents offered this answer as an example of "other" entities that should be held responsible). A common metaphor was that each person should "put their little grain of sand on the pile, then together we can all help everything grow."[29] The remainder were unsure of how to assign or divide responsibility. At least one person (the husband of an Alta Gracia worker) referred to the company as "helping people. . . . They've helped people, but within the same firm. People have had setbacks and the same firm will help them," but in his opinion, well-being was ultimately "up to the person himself."[30] Capturing the idea of subjective socioeconomic status (SSS) explicitly, one young mother (herself an independent artisan), when asked about living conditions in her community, responded: "For me, the level of well-being of people in Villa Altagracia is changing. . . . It's low, but it's changing—it's all relative, because there are people who have a lot of money, who live well, and others who live supremely poorly. I see it all as dependent on the type of person we're talking about."[31]

GRASSROOTS PERCEPTIONS OF CONSTRAINTS ON STAKEHOLDER DIALOGUE

The 10-question survey instrument designed to guide interviews in Villa Altagracia and Bonao alike progressed from an opening life history, through questions on a respondent's perceptions of community well-being and employment opportunities, to a block of questions related to how respondents glean information on corporate practices in their community.

When asked how they obtained information about "company policies or practices in your community" and given a list of response categories (ranging from personal connections to formal media to social media to company materials to "no information" or "don't know"), respondents in Villa Altagracia answered in a variety of ways. Of the 27 persons interviewed, almost half answered that they used social media or conventional media (i.e., six used social media; six used radio or television), but those who use social media were all either college-aged people or young parents whose occupations ranged from homemakers and informal artisans to truck drivers to a nail salon owner. Eight people responded that having personal connections who could share information on hiring by local firms was the only way to obtain information on corporate activities in the community. Two other respondents answered that they obtained information through representatives of companies themselves. (For example, a small hardware store owner explained that he would speak with sales representatives to obtain new product information; a former employee of the local citrus cooperative explained how commercial marketing of products was used to share information with consumers.) At least six people interviewed responded that they had no information on corporate activity because it was scarce and difficult to obtain.

The non-random nature of my sample meant that many people had heard of the Alta Gracia Apparel company—not surprising in a town that is the namesake of the Alta Gracia "project" (as several of the Alta Gracia workers and their families referred to the company, along with Adler-Milstein and Kline 2017, 54)—but even they were less familiar with the specific policies of this firm. Nor were many familiar with the policies of other firms in town. Because there was limited publicly available information on company practices in their community, residents typically found out about hiring by word of mouth. Several argued that the local news was effectively controlled by larger corporate interests in Santo Domingo and elsewhere, decreasing the likelihood that the truly challenging circumstance of people in Villa Altagracia would make the news or that corporations' activities would be accurately reported on.

As the owner of the tire repair business explained: "Well, you don't learn anything from companies here. . . . You're the first person who's ever come to this neighborhood to interview me in relation to my work—and I've been on this corner for a long time."[32] One of the former BJ&B workers I interviewed, when asked how she obtained information about corporate practices, answered: "It'd have to be through gossip, because the companies themselves are really slow to share information. They don't say, 'Hey, there's work in this or that area; come on and apply, workers.'"[33] Among the people I interviewed with a higher level of skills and better job prospects, such as the son of an Alta

Gracia worker (who had taken several technical courses and was working in a skilled trade), information on job opportunities was also typically passed from insider to insider: "I do have a friend who works in a good company, and we share ideas about who's hiring, how it's going, who earns more and how much. So, we share ideas. . . . It's hard to get this type of information through social media or over the news."[34]

ASSESSING LOCAL KNOWLEDGE OF AND RECEPTIVITY TO STAKEHOLDER DIALOGUE

Only after establishing a foundation of life history, normative expectations, and knowledge of corporate practices could I then probe people's understanding of stakeholder dialogue practices explicitly. I asked respondents if they had ever participated in (or at least heard of) company–community consultation or would be willing to take part in such a process. Three questions enabled me to gauge awareness of and receptivity toward stakeholder practices, including a factually based question (i.e., "Occasionally, people from the community meet with company representatives. Has that ever happened in the community where you live? Tell me about it."); a hypothetical question ("If you were invited to participate in the dialogue with representatives of the company, would you participate? Yes or no and why?"); and a conditional question ("If you did participate, then: a. what would your top priority be? b. what kind of information would you need to make the conversation effective? and c. what kinds of resources or skills would you need to make the conversation effective?" See Appendices 1 and 2).

The answers to these questions reveal a range of complex factors that complicate the landscape for stakeholder consultation. Beyond the two Alta Gracia workers I interviewed, no one else had ever participated in any type of company–community dialogue. As a bus driver explained: "The companies here only do meetings with their workers, never with the community. Not with neighborhood associations, nothing. . . . [It's] at the company's convenience, but this has never happened in Villa Altagracia. We'd go, but they never invite anyone."[35] Some indicated they viewed stakeholder consultation as a means to a concrete end and would participate accordingly. As one college-aged man explained: "When we're talking about participating I'd have to see what the dialogue focuses on—there might be something nice that could come of it. But if I don't see that potential then I'd keep quiet."[36]

The variation in an individual's level of willingness to take part in consultations depended in large part on the stake that person believed she or he had in the outcomes of collective action. Some were willing to take part in consultation even if the process itself remained somewhat unclear. The truck driver I interviewed was originally from a different hometown but now spends a good deal of time in Villa Altagracia because he has an infant son in the community. When asked whether he would participate in dialogue with a local company, he replied: "Sure, because it would be a way to give support to the people of this town and improve future well-being—one has to cooperate."[37]

For others, however, a sense of fatalism outweighed their willingness to risk engagement with companies. This was the case with two grandmothers (one middle-aged and another in her 70s), who responded to the hypothetical invitation to participate in stakeholder dialogue with a definitive "no." The younger of the two explained: "I wouldn't go because I don't like the conclusions. I don't take part . . . in these things. I'm more conformist. We have to solve our own problems; we resolve to do it—try to survive as God helps us."[38] The other argued that she thought it best to "conform to what God has given you, because you cannot expect more."[39]

When asked conditionally what they would need in order to make the process effective, many of those I spoke to who were willing to engage in consultation offered detailed responses. One of the shop clerks I interviewed explained, "I'd like at least some kind of training or some way to learn more about people's needs,"[40] while the college-aged son of another Alta Gracia worker explained that he would need "information" if, for example, a company were polluting a local waterway: "I'd need to know where the river is, what's happening to the river. . . . My main priority would be to be informed."[41]

An Alta Gracia worker herself observed, however, that many of her neighbors do not think about the type of information they would need to make dialogue with a company successful: "Sadly, we're more interested in *telenovelas* (nighttime television dramas), in sitting there watching nothing" than in learning about community issues.[42] Indeed, the wife of another Alta Gracia worker I spoke to plainly acknowledged that she had little information on corporate activities in her community because "I don't watch the news, I watch *telenovelas.*"[43]

Among the best-informed respondents—including an Alta Gracia worker who had taken an active role in the formation of the union in the plant—there was a perception that "community meetings have happened here before" but other "companies didn't come. They were invited with letters and everything. They didn't come ... Decisions were taken, and when we're all together up front, things look very beautiful. But later, no one complies. ... Especially in the [citrus processing] company" on the outskirts of town "where people buy these juices—this business doesn't comply with anything."[44]

COMMUNITY ATTITUDES ON THE NATURE OF REMEDY

Finally, I asked respondents whether they agreed or disagreed with the following statements, all of which relate to the nature of remedy, on some level:

   a. companies do not have a social obligation to people beyond their employees or shareholders;
   b. if a company's leadership chooses to make a contribution to the community (such as support for a local school or clinic), community members should accept it and not expect more;
   c. labor unions can only help their members, not people outside the union.

Responses were evenly split between those who believed that companies do not have responsibilities to people beyond their own employees and those who believed they do (i.e., 12 to 12, with the remainder "don't know"). There was a near-even split among those who believed unions do not have responsibility for anyone beyond their members (i.e., 11 agreed, nine disagreed), but nearly as many people argued that they themselves simply didn't know enough about unions to answer this question (i.e., seven people). This high rate of people who "don't know" enough about unions to feel competent to answer reflects the low unionization rates in the Dominican Republic more generally, as discussed earlier in the chapter.

Respondents were slightly more likely to believe that if the company chose to make a contribution to the community (discretionary philanthropy), then community members should accept it and not expect more: 14 people agreed with the statement, 11 disagreed, and the remainder answered "don't know." In part, the idea that members of the community should be grateful for whatever was offered may have been spurred by the idea that they were unlikely to receive social support through government or other channels.

However, many respondents thought corporate earnings or benefits should extend only to employees; anything a company chose to give above and beyond that was voluntary. As the bus driver I interviewed explained, if he were invited to a consultation with a company, "I'd need to ask them to better regulate electricity here, [to deal] with the problem of lack of potable water . . . [I'd ask] that they construct houses for the poorest of the poor—because they can, they're huge companies. . . . [But] the company itself isn't obligated [to do this]. Whatever they give is voluntary."[45] The majority of respondents tended to be unaware of any type of compliance requirements relative to positive social obligations on the part of the company. The notion that a company had a responsibility to remedy broader social ills was foreign to nearly all respondents in Villa Altagracia. Since the bulk of factories left the main freetrade zone close to a decade ago, the overarching concern in town was how to replace the thousands of jobs that had dried up when the plants had closed, not to make demands on the existing companies in the zone.

LESSONS LEARNED

Four key points emerge from the interviews in Villa Altagracia that require us to think differently about the nature of stakeholder dialogue and the challenges of transcending the current limits of the practice.

First, even in the town where a model factory has forged a new model of worker-driven social responsibility for corporations, there is nevertheless a widespread lack of information at the local level about company practices or the notion of stakeholder dialogue more generally. The messaging regarding Alta Gracia, the building of its brand, and its corresponding business model have all been outwardly oriented—toward stakeholders in the consumer apparel markets where university apparel is sold and worn outside the Dominican Republic, rather than toward people at the local level within the community itself or at the national level. The factory's business model isn't well known beyond the people directly employed or their family members or acquaintances. This, in turn, lessens the likelihood that consultation is understood or expected vis-à-vis the Alta Gracia company or other companies in the town.

My participant observation with people involved in the NGO community more generally yielded similar insights.

People involved in community development work had never heard of the triple-wage factory, whether they worked for large international development organizations with offices in Santo Domingo or more local groups involved in basic health, sanitation, or educational promotion in hamlets just on the outskirts of Villa Altagracia. Even people involved with local NGOs that routinely hosted US college students as volunteers for summer or spring break construction work were unfamiliar with the company or this business model, as were local hostel owners in the capital who routinely hosted Peace Corps workers, missionaries, and other volunteers in the helping professions. Government representatives from whom I gathered data were aware of the general development context in both towns, and one representative was familiar with a union organizer involved in the Alta Gracia factory, but none of the government agency staff I encountered seemed aware of the specific living wage factory model.

Second, most non-workers in Villa Altagracia do not view stakeholder consultation as a practice that they have a "right" to take part in, nor do they expect that companies or labor unions "owe" them any concrete remedy for harm of any sort, since they lack an employment relationship with Alta Gracia Apparel or other firms. The people I interviewed did, however, tend to believe that they have a right to good governance and to social welfare benefits allocated through the state. Because many believe that state actors can attract companies to their town, they argued that unemployment is a function of government's lack of capacity or will to induce companies to create jobs (rather than viewing companies as part of intensely competitive, globally driven industries). Only one person argued that factories left Villa Altagracia because of competitive pressure—and in this case, the argument was that taxes in the Dominican Republic were too high, so companies like BJ&B left (not that they left in search of lower-wage countries in which to manufacture). This same respondent (a bus driver) did argue that a company had an obligation to tell workers "before it leaves . . . If they are going to leave, they have to give each worker severance pay, because that's a law." But he did not argue that such a right to information about factory closings would extend to other people in the community more broadly.[46]

Third, many of the people I interviewed in Villa Altagracia were caught up in the struggle for daily living and appeared to have neither the time nor resources to spend in consultation with companies if the exchange was not connected to concrete benefits. None expressed direct envy of their friends' and neighbors' triple minimum wage jobs (though social desirability bias could explain the reluctance to do so explicitly) (Callegaro 2008, 826). But several noted that companies often use public information to advance their own market share, not to advance community interests. As one woman who had worked for the citrus consortium on the outskirts of Villa Altagracia explained, companies "communicate about themselves through posters . . . Here they do a lot of promotion about donations for campaigns against cancer, or donations to the poor" through product-based labeling. "Not all of their products" are marketed this way, she acknowledged, but some are, "and the public understands this type of marketing."[47]

The notion that the benefits of stakeholder consultation would likely be higher for companies than community members involved in consultation was linked, in part, to the broader pessimism many respondents expressed about the willingness of government representatives or others in positions of authority to truly empathize with or care about the fate of poor people. Visits to poor neighborhoods or offers to collect résumés were all seen as empty gestures by government representatives—and this cynicism often extended to the idea of stakeholder consultation.

Fourth, community members have creative ideas about how best to use the process of corporate consultation to advance collective well-being, were they to be involved. In part because they believe that corporations have significant resources, respondents often tended to assume that companies could provide "big" benefits to the community if they chose to, such as better housing[48] or infrastructure improvements, or more jobs. When asked if they would participate in stakeholder dialogue, Alta Gracia workers themselves or their family members seemed most predisposed to do so. As one worker explained, "Yes, of course [I would participate] . . . because I'd like to do battle about unemployment and call on [companies] to create more jobs; that's something we really need."[49]

But another Alta Gracia worker was quick to note that charity without consultation was not necessarily helpful: "I've always critiqued donations [given] without creating the means to earn—just giving things. When some people come from the USA or Europe, or even some local Dominicans, and want to make a donation to a town or community or a family, they come with a packet. They bring four pairs of shoes, and I think, 'I don't need a pair of shoes; what I need is a workbook and a pencil.'" Paraphrasing the thoughts of other friends and neighbors who encounter this type of charity, she explained that many would likely reply: "'What I need is your help in making it possible for my daughter to go to high school, because she's currently not able to go to high school.' So, I like much better the people who come and say, 'What are the things that the people in this neighborhood need, or this community or this family?' Come first and lend an ear. And then afterwards, you can help."[50]

There is already precedent for people in neighborhood associations to meet with "government functionaries," a domestic worker I interviewed explained: "There are neighborhood associations that get together when transformers are damaged and there are power cuts. The neighborhood associations and groups get together and they try to resolve things."[51] But whether or not these would provide a platform for consultation was not clear in most interviews. A retired garment worker noted the existence of neighborhood associations but did not make the jump to seeing them as a vehicle for stakeholder consultation: "Here there's a neighborhood association. . . . But I really don't know what it does . . . I went and participated, I voted, but I am not sure what they accomplished."[52]

Collective action was most often noted as a form of making demands on government representatives, not companies. The son of an Alta Gracia worker, for example, recounted:

> In this neighborhood there wasn't a bridge (the one that you took to get here wasn't here) and a lot of people were hit by cars. Weekly, people were hit and died. So, the people had a strike—in other countries they call it a protest, but here we call them strikes, where people burn tires, they burn trash, so the transit stops and there's no way to get around. And

for that reason, one supposes, they [the government] built the bridge.⁵³

Workers from the Alta Gracia factory themselves have had experiences of union leadership within the plant, which left them realistic about the challenges of collective action but also hopeful. In order to make the process of stakeholder consultation effective, one worker argued:

> I'd like first and foremost to have the opportunity to get together with a big group of people to find out what they think about issues, what do they want? And to bring them information, too: we don't only want electricity . . . we're also going to think about what we *do* with that energy. What is more convenient for you—energy? Or fixing up the schools? Which do you want more: fixing up the schools, or fixing up streets? These are all things we have to find out through conversation with various people, who feel very comfortable thinking about these competing priorities and could say, "I have five children; I don't have access to the Internet. Is it better for me if they put together a computer lab in the school . . . or is it better if they arrange my street beautifully?"

She said that "sitting down with people who can think freely and decide completely and correctly for everyone, not only for myself" was central to her vision of how to increase the chances of successful stakeholder consultation.⁵⁴

One of the most creative visions for Villa Altagracia's future came from a young artist and father who did not look to conventional manufacturing companies to provide either the platform for development or the vision for the type of economy that would best suit the town moving forward. He envisioned an alternative economy largely dominated by the arts and services in the wake of industrial decline:

> Political actors—to whom we often assign responsibility for community development, principally—are very conflictive among themselves, so they don't achieve unity and don't achieve long-term plans for development, not even short-term community development . . . Art per se doesn't have a headquarters specifically. . . . You could say that Villa Altagracia is artistically virgin terrain, in the sense of a market for publicity, events production, and for music, too. In terms of entrepreneurial development . . . we are proceeding with a form of development that, although it'll be very long term, will create more demand for this type of labor. . . . We use Santo Domingo as a jumping-off point for reaching other markets [such as those online]. . . . So, we're positioning ourselves for when development happens. . . . With the theme of globalization, we understand how the world's development [happens] and we want a life like that. So there are a lot of people who want another type of life, and although we don't have it, at some point, it'll develop and we'll achieve it. . . . We haven't taken advantage of all the potential niches of the market in the Altagracia.[55]

IMPLICATIONS

So, what do these divergent life experiences and visions yield in terms of the prospects for stakeholder consultation in Villa Altagracia? *First and foremost, a view from beyond the factory reveals the complexity of the challenges people face on an everyday basis.* Problems of infrastructure—basic things like electricity and water—often overwhelmed peoples' capacity to take time out to participate in a formal consultative setting (even if the potential for doing so was only hypothetical). *Second, worker-driven modes of consultation have proven successful in places like the Alta Gracia factory in achieving living wages for several hundred workers whose product is highly visible internationally. But the dearth of employment opportunities beyond the factory swamps these gains for most people in the community.* To address the root causes of community concerns revealed in my interviews, stakeholder consultation would need to involve a much broader range of actors than it currently does. But there are not widespread mechanisms for doing so here or in most manufacturing communities globally, as Chapter 3 reveals.

Third, community members often have innovative and interesting ideas of how to reimagine their economy (like that sketched by the artist at the end of this chapter), but there is not a vehicle for integrating non-workers into consultative practices in most conventional multi-stakeholder initiatives (Collins, Evans, Hung, and Katzenstein, 2017). The worker-driven social responsibility model at Alta Gracia Apparel offers an alternative, but direct community engagement is still limited in scope and coverage.

Companies could translate ideas like those that emerged through my interviews with local residents (such as having companies help address the community's chronic lack of access to electricity) in practical and targeted ways, such as putting solar panels on a factory and then sharing energy with key public buildings in the local community (e.g., school, health clinic, or community center). Other ideas, like opening a computer lab for public use after hours (already proposed by Alta Gracia workers as part of their collective bargaining process with management and as referenced in at least one interview), would have impact not only on digital literacy but also on new business development (i.e., if local people with start-up ideas for new income-generating ventures had access to online resources). A two-way dialogue between companies and communities would transcend the charity model of engagement and likely produce more useful and appropriate modes of engagement on the part of the company.

But without government involvement, many of these novel ideas would not be sustainable; indeed, the *fourth key implication stemming from the Villa Altagracia field interviews was that government has a credibility gap at the grassroots level* (consistently referenced in interviews). *Thus, stakeholder dialogue needs to engage government more centrally as a party jointly responsible for meting out remedy* for problems in manufacturing communities. The relationship between companies and government needs to be clearly specified in stakeholder consultation. Existing research on multi-stakeholder initiatives acknowledges government's role (Collins et al. 2017), but dialogue in practice is largely vertically organized, with companies and government interests being more aligned than either party is with community members (despite a public mandate to govern in the public interest). Interviews in Villa Altagracia indicate that people at the grassroots level often view government and business as partners in determining social welfare outcomes in their community—but are keenly aware of a lack of political will to govern transparently, equitably, or efficiently.

Fifth and finally, communities themselves are not homogenous and this, in turn, affects individual people's willingness and ability to take part in consultation (even when the possibility of doing so is only presented hypothetically). In my interviews, people's openness to stakeholder dialogue ranged widely: from those eager to try, to others who believed fatalistically that they had little to no ability to influence outcomes and so would not bother to take part. SSS influenced individual people's receptivity to consultation, with better-educated or better-off respondents on average expressing a willingness to engage in consultation. Individual perceptions of SSS varied, however, with some deeply constrained by illiteracy, sick family members, or single parenthood and others visibly disabled yet guardedly optimistic about their own prospects. The diverse views about who is responsible for community well-being and how best to go about organizing to achieve it were evident in the voices of people heard in this chapter. Together, they reveal the challenge of determining who is inside or outside of the consultative process and the limits of determining the scope, allocation, and content of remedy. The following chapter takes up the case of people in the neighborhood of Las Delicias in Bonao in order to explore whether a different approach to textile manufacturing in that community shapes the context for stakeholder dialogue in ways distinct from Villa Altagracia.

CHAPTER 5

Challenges Down the Road

Stakeholder Perceptions in Bonao

There really is no one taking responsibility for what happens here. . . . It's all about who's doing favors for whom; you can't get anything if you don't know someone.

—Retirement-aged homemaker[1]

The biggest thing that worries me about this community [is] when someone . . . wants to do something, the majority of people want to block the way. . . . It has gotten to the point that—how can I tell you? The mandate is "Me, always me." We have people who, for example, put on events but do it more for show outside than to help people inside.

—Woman farmer and grassroots organizer[2]

BONAO IS SLIGHTLY NORTH OF Villa Altagracia, 40 kilometers farther away from Santo Domingo. Located in the state of Monseñor Nouel, in the foothills of the country's central mountain range, Bonao's name may not be synonymous with a particular brand of clothing (as is Villa Altagracia), but it also hosts a key player in the textile industry, Hanes Brands. Hanes operates a major spinning factory in Bonao's Dos Rios factory zone, employing upwards of 3,000 workers who produce bolts of fabric in four shifts, 24 hours a day, seven days a week. These workers make up roughly 2.4% of the population of just over 125,000 in Bonao (in contrast, the 200 Alta Gracia workers constitute roughly 0.23% of Villa Altagracia's population).[3] Bonao is one of three Dominican towns where Hanes has produced textiles for over 30 years. With over 67,000 employees in more than 40 countries worldwide, Hanes had net sales of over US$6 billion in 2016 (Hanes Brands 2017, 11, F53). The plant in Bonao is a conventional manufacturing operation, paying standard wages. In 2015, Hanes Brands acquired Knights Apparel, the original owner of Alta Gracia Apparel, although Alta Gracia was not part of the sale and instead became an independently owned and operated company (Adler-Milstein and Kline 2017, 132–133; BusinessWire 2015). So, less than an hour apart, two distinct approaches to textile manufacturing continue to unfold in the Dominican Republic.

This chapter explores local community members' perceptions of stakeholder issues in Bonao, through interviews with people in the neighborhood of Las Delicias. Bonao's economic base includes not only manufacturing but also natural resources (e.g., nickel, bauxite, gold). The Canadian mining corporation Falconbridge has been active in nickel extraction and smelting in and around Bonao since the mid-20th century, undergoing various corporate mergers and acquisitions well into the early 2000s. As recently as 2015, the subsidiary of Americano Nickel Ltd., Falconbridge Dominicana SA ("Falcondo"), aimed to mine the nearby Loma Miranda mountain (BN Americas 2018), but not without considerable community opposition (Salgado 2016). Another Canadian extractive company, Barrick Gold, faces ongoing community protests just

30 kilometers northwest of Bonao over the environmental impact of its activities. Although people throughout this region have engaged in collective protests focused on health problems stemming from mining activity (*The Economist* 2013), none of those I interviewed in Las Delicias in early June 2017 had done so.[4] An older man who had worked as a security guard for Falconbridge and related companies since the mid-1990s characterized it as "good work," pointing out that he earned roughly $210 a month—a third less than a typical Alta Gracia worker would.[5]

Because the extractive sector is already over-researched in the literature on stakeholder dialogue (as demonstrated in Chapter 3), this chapter explores instead whether the presence of light manufacturing companies with differing business models in the two towns (i.e., Alta Gracia in Villa Altagracia and Hanes in Bonao) leads to variation in local people's knowledge and attitudes regarding stakeholder dialogue. Garment manufacturing in the Dominican Republic accounts for a larger segment of the economy than extractives. The extractive industry plays a limited role in generating gross domestic product (GDP) or employment in many countries of Latin America—including the Dominican Republic, where extractives accounted for less than 2% of Dominican GDP in 2013 and just 3% of employment on average across most countries of Latin America in the same year (Helwege 2015, 74–75).[6]

This chapter extends the discussion of the prospects for stakeholder dialogue in light manufacturing through the interviews I conducted in the relatively low-income neighborhood of Las Delicias (selected in order to facilitate comparison with Villa Altagracia) using the same questionnaire to explore how respondents' subjective socioeconomic status (SSS) affects their perceptions of responsibility for community well-being and their knowledge of and willingness to participate in stakeholder dialogue. The interview sample in Bonao was again a snowball sample, built through contacts initially shared by a former development worker. In Bonao, as in Villa Altagracia, I did not interview company management, nor did I conduct interviews with workers on site at the manufacturing plant. The one Hanes worker interviewed in Bonao was a young man I met while walking through the community.

Both towns are situated along the Dominican Republic's Highway #1, the spine of the country's major industrial corridor. The highway connects Santo Domingo on the southern coast with the manufacturing hub of La Vega and then extends northward to the coastal region near Puerto Plata, one of the country's first major tourist resort destinations. The highway marks the daily lives of people in both towns: it is loud, large, and dangerous (in terms of not only speed and congestion but also crime). It is also the transit route for fabrics produced in Bonao that are cut, assembled, and sewn into products at other plants throughout the country. As the young Hanes employee explained while we talked outside his family's home, "We make the fabric," at the factory in the Dos Rios manufacturing zone in Bonao, so that other workers elsewhere "only have to put together the collar, for example, or the bands on the arms in another factory. This means that you pass the product from plant to plant—collar to band—and you end up with the finished product."[7] As these products move up and down Highway #1 and beyond, do the prospects for stakeholder dialogue change in the various communities along the way? Why or why not?

EVIDENCE OF EVERYDAY CHALLENGES

The neighborhood of Las Delicias is divided from the rest of Bonao by the highway, and people in the neighborhood often pointed out that on their side, there is no grocery store, no medical clinic, and no high school. The din of multiple lanes of traffic is a constant in the background, and one woman (a homemaker with young children) remarked that people are killed or hurt time and time again in Las Delicias, trying to cross the highway simply to carry out basic errands or to get to work.[8] Even the major thoroughfare within the neighborhood was only recently paved—and only after a community strike and violent protests in 2015.[9]

Las Delicias is a poor neighborhood in comparison with those elsewhere in Bonao, as one woman farmer and grassroots organizer who lived on the outskirts of the area explained: "Let me tell you, when you go down here, you'll find people who, by three o'clock in the afternoon, don't know where their next meal is coming from. Down here, there are people who don't know when they are going to put on more clothing. Down here, there are children who go to school in flip-flops and if they don't have them, their neighbors have to step in to supply it—I'm telling you because I have had to help a lot and if I can't, then I know where to ask to get it from on the other side" of the highway.[10]

In Bonao, the 3,000 workers employed in four around-the-clock shifts by Hanes Brands do not work in a "living wage" factory: The legal minimum wage is the Dominican peso equivalent of 83 US cents per hour versus the Alta Gracia wage of $2.83 an hour, though take-home pay varies depending on overtime and other factors. As respondents often remarked, jobs in the textile industry of Bonao are not jobs that will make you rich, but they are jobs that open the path to some measure of economic security, albeit limited. While at least one of my respondents reported making the nearly four-hour roundtrip commute from Bonao to Santo Domingo, this person did so for a higher-paying job in finance. There was less evidence of daily or weekly commuting for the type of domestic service work or day labor in construction more common among people in Villa Altagracia—in part because the capital is roughly two hours away. During participant observation in Las Delicias in Bonao, I observed two homes where children were being watched by neighbors or by extended family members in the absence of parents who worked out of town for several days a week or were entirely absent. But commuting to the capital was markedly less referenced in my conversations with people in Bonao than in Villa Altagracia.

Common to both places were significant problems with infrastructure, particularly electricity. In Las Delicias, just as in Villa Altagracia, power outages were a routine feature of life; only the scheduling was different, with alternating days of morning and late afternoon outages as opposed to nightly outages in Villa Altagracia. This meant that families and people who ran local businesses or directed institutions like the local orphanage[11] all adapted their routines to the anticipated outages. Also common in both communities was the high rate of teen pregnancy and the vulnerability of young women to the pressure of older men (Dominican as well as foreign) who seek them out for companionship, creating difficult tradeoffs for women in economically strained circumstances.[12] In Bonao, several of the women I interviewed were as young as 13 or 14 years old when they initially entered into common-law relationships. One explained that following domestic abuse by her brother, "I said, 'I'm getting out of here!' And I got married because my brother was beating me."[13] In the neighborhood of Las Delicias, efforts by local religious organizations to create teen fellowship groups as well as parenting clubs have drawn upwards of 50 people (adolescents as well as parents) into regular activities, as revealed through participant observation. A former development worker for the community similarly sought to create a parenting support network in Las Delicias in the past.

But linking parents and youth to opportunities, resources, or support outside the neighborhood remains a challenge—one marked, in part, by the cost of transportation and the difficulty of sustaining involvement when families are balancing economic pressure with concern over crime (which, for some, was a deterrent to attending evening activities). While Bonao does not have as high a crime rate as Villa Altagracia, a retirement-aged homemaker nevertheless observed, "Here they're killing lots of people, attacking them, there's a lot of delinquency—too much. One can't even go out; from 6 p.m. onward, you can't walk around. If you go to the other side of the highway, you have to go in a taxi or car. On motor scooter, no one dares to come, because they could be attacked."[14] Indeed, the frequent references to concerns about personal security were a common theme in both towns and shaped decision-making, particularly among women.

GLOBAL PRESSURES, LOCAL REALITIES: STEPS BACKWARD AND FORWARD ON BUSINESS AND HUMAN RIGHTS

In Las Delicias, it was common late in the afternoon, just before the start of second shift, to see workers walking out to Highway #1 to catch a bus to the Dos Rios factory zone or riding motorcycles or scooters in tandem. They all wore red or blue Hanes polo shirts or T-shirts embroidered with the company name. Hanes products (like the shirts the workers wore) are integral to collegiate apparel production, though not made exclusively for this market. Other plants that specialize in cut-and-sew operations (like Alta Gracia Apparel) can use Hanes fabrics in licensed production of college-logoed apparel. Factories that specialize only in screen-printing or embroidery work can also use Hanes blanks (i.e., undecorated shirts or other garments) in the production of collegiate apparel.

The Hanes plant at the Dos Rios factory zone in Bonao has been the site of labor unrest at various points over the past three decades, coming to a head in 2006 to 2008, with the Workers' Rights Consortium reporting on extensive grievances related to freedom of association, overtime pay, benefits, and basic treatment of workers (Workers' Rights Consortium 2007; 2008). The Dominican Labor Ministry was also involved in investigating complaints during this period (Workers' Rights Consortium 2008, 2). Although managers at Dos Rios have adjusted overtime wage and hour practices along with safety procedures, the plant has also refined a variety of lean manufacturing techniques. So, Hanes would thus seem typical of standard problems in the light manufacturing sector, where changes in labor rights are contingent on struggle and production innovation is tied to becoming leaner, not to a living wage.

But this case departs to some extent from the standard story given the nature of corporate philanthropy carried out by Hanes in Bonao. Roughly at the same time it was involved in this labor unrest, Hanes forged a partnership with the local Hospital Pedro E. Marchena in Bonao: Hanes staff initially volunteered to renovate the pediatric and maternity wards, donating hours on company workdays. Beginning in 2012, the company began to support an extensive screening program for pediatric ear, nose, and throat problems and to underwrite the cost of free surgeries for 50 children annually in communities where it has operations across the island. Using a mobile health unit (provided by the company to Marchena Hospital), local staff and medical volunteers from Wake Forest Baptist Medical Center (in North Carolina, where Hanes is headquartered) have screened children at health fairs in those communities. Children identified in the annual cohort then have free surgeries, carried out at Marchena Hospital. In a few more complex cases, Hanes Brands has underwritten the cost of transportation to and surgery at Wake Forest Baptist Hospital. Nurses and residents from Marchena Hospital have since taken part in a two- to four-week training immersion program at the North Carolina hospital. The collaboration is organized at Hanes headquarters in Winston-Salem, NC, by a Dominican-born staff member and managed at the Dos Rios plant by a local counterpart who operates as a liaison to local families involved, according to respondents I interviewed whose children had received free surgeries through the program (BusinessWire 2012).[15]

Yet beyond these families, my interviews in Las Delicias revealed a relative lack of overall awareness of Hanes's medical philanthropy program. Only a quarter of respondents (i.e., four out of 16) had heard about this activity, and often only vaguely. As the owner of a small corner store (*colmado*) explained: "I've heard about it, but I don't know where it's happening . . . I've heard that companies in the zone help a lot with what people need. And I've heard that they've done operations," but his general lack of specific knowledge about the activities, coupled with a tendency to conflate the charitable activities of various companies, was common.[16]

FINDING WORK, BUILDING SKILLS: NETWORKS AS GATEKEEPERS

Respondents were much more sharply attuned to how employment opportunities at Hanes circulated through community networks, or what training opportunities workers or others in the community could receive through the company's ongoing collaboration with the Dominican Government's Ministry of Labor. Sociologist Andrew Schrank (2013) has written extensively on the Dominican Republic's efforts to professionalize its Ministry of Labor in the early 2000s, largely in response to threats that the United States would withdraw preferential trade access if the country did not address longstanding labor rights abuses in manufacturing. (Recall from Chapter 4 that in the 1990s, the Dominican Republic was one of the top exporters of textiles to the United States.)

In response, the Labor Ministry created a specialized institute on professional training (Instituto Nacional de Formación Técnico Profesional [INFOTEP]) within the ministry, initially to improve the skills of its own inspectors. Over time, INFOTEP has developed a wide range of vocational training courses as well as management training and certification programs aimed at improving the employability of people and the productivity of businesses across a range of industries and trades. In 2016, Hanes-Dos Rios collaborated with INFOTEP on 88 different training activities, involving 1,888 participants (both workers and community members from various neighborhoods in Bonao) in over 3,600 hours of courses.[17]

In Bonao, just as in Villa Altagracia, my interviews included community members who had participated in INFOTEP training courses or who knew friends or family members who had done so. These respondents tended to view INFOTEP courses as a route to professional mobility. Whereas the INFOTEP course graduates I interviewed in Villa Altagracia had taken part in training for skilled trades (e.g., soldering or electrical work), the majority of those I spoke with in Las Delicias in Bonao had taken more basic courses (e.g., those necessary to complete a general equivalency diploma for high school or life skills courses on conflict management). The orphanage in Las Delicias had also hosted workshops on interpersonal communication for its staff and people in the community, facilitated by trainers from INFOTEP with support from Hanes.[18]

People who had taken part in INFOTEP training programs commented on the quality of instruction, whether they had taken Saturday classes to complete a high-school equivalency degree, a one-day workshop on communications, or months of classes toward certification in skilled trades. Some credited their INFOTEP certificates with making them more attractive to employers. As the young male employee of Hanes I interviewed argued: "Look, one who has a course in INFOTEP has a job for sure. Because if [they try to turn you away and] you say, 'No, I have a course in INFOTEP,' all at once they'll say, 'Bring your résumé,' because INFOTEP is based on real learning."[19] He, like others, commented on the free transportation and lunch attendees received if they took employer-subsidized courses on a weekend.

In Bonao as in Villa Altagracia, every person I interviewed who had been trained by INFOTEP in a skilled trade was male; all considered themselves comparably better positioned in the labor market than other people in their communities. Even amidst the worsening competitive state of the Dominican textile sector, Schrank argues (2013), blending labor rights compliance with the promotion of broader skills development (as INFOTEP has done) has helped slow the race-to-the-bottom scenario characteristic of the global textile sector.[20] Worker salaries may be uniformly stagnant or even lower in real terms in the Dominican textile sector than at other points in time, but access to training through INFOTEP has become integral to some firms' (such as Hanes's) approach to social responsibility.

But there is a deeper reservoir of economic pain as well as entrepreneurial potential that conventional industrial training efforts cannot address—both because of the gender gap in training and skills development and because existing models of labor rights promotion and stakeholder dialogue miss the broader problems of the community contexts in which manufacturing takes place. The Bonao case illustrates the internal contradictions in the business and human rights (BHR) field and the challenge of determining when remedy begins and ends. On the one hand, Hanes was the focus of investigations by the Workers' Rights Consortium related to violations of basic labor rights in the early 2000s. On the other, its ongoing philanthropic efforts demonstrate a high level of innovation and some degree of year-on-year sustainability.

One challenge is to reconcile the two faces of BHR—namely, the shop floor/workforce face versus the one that community members see. Another challenge is to place BHR efforts within the context of broader challenges of underdevelopment and underemployment in towns where manufacturing takes place. As people I interviewed in Las Delicias were quick to argue, even a relatively low-paying job was better than nothing. The bar for BHR here is thus extremely low.

THE INFLUENCE OF SUBJECTIVE SOCIOECONOMIC STATUS ON ATTITUDES AND ACTION

The low bar is a result, in part, of the extremely limited employment and income-generating options available to many people even in manufacturing towns like Bonao, where the main textile factory (Hanes) is operating at full capacity. A childcare worker in Las Delicias used her brother's experience to illustrate the potential and limits of working in the textile industry in Bonao. Her brother had worked for Hanes for five months at the time of our interview:

> He has a high school diploma. He had two semesters in college . . . studying accounting before leaving to work at Dos Rios . . . [they] give you the opportunity to study . . . The courses that they offer are ones you take on your day off . . . The company reimburses [course costs], so the people can pay for their studies. And it is for this reason that the majority of young people want to enter Dos Rios—because they show you a lot of opportunities. They prepare you as a technician and they also prepare you with an academic career. . . . Imagine the opportunities that you have: you have medical insurance, you have a salary—not a "wow" one, not a luxurious salary, but at least an economic basis from which you can live comfortably. I'm not saying that it's a luxurious salary, but it's better than doing nothing; it's good.[21]

The limited employment opportunities in her community made this kind of work—albeit at a significantly lower salary than Alta Gracia—better than "standing around idle in the street, without anything to do . . . There are two or three young people who are working in Dos Rios" in her neighborhood, the childcare worker explained, "but there are lots of young people now who we've asked Dos Rios to help—and we want to see if through the *junta de vecinos*" (neighborhood association) there might be other positions available. "*Aye caramba*, there are more than 50 young people" here without work, she exclaimed. "It's a disaster."[22] As a woman farmer explained: "The majority of young people in this community walk around goofing off, driving around on motorcycles from house to house. Those who work are there," in the Dos Rios plant. "For adults, it's the factory zone (*la zona*), there's not much else. . . . For the majority of women, it's hard to get work, because they haven't studied, so that makes it harder for them."[23]

People with better employment opportunities and income tended, not surprisingly, to assess the economic situation in both towns (Villa Altagracia and Bonao alike) as stable, albeit constrained by the deeply personal nature of how work opportunities are allocated. This sentiment was best captured in Bonao by a man who worked in finance. As he explained: "Things maintain themselves sort of stably; things go up and down, but are relatively stable, you understand? But they don't get better overall," though in theory they should "because here we have Barrick Gold, Falconbridge, the free-trade zone of Dos Rios, we have another old free-trade zone (no one knows who's in that zone!). So there's not a lack of industry—if all these enterprises could employ all the qualified people. But they don't."

Reflecting further on why unemployment and underemployment persist in Las Delicias, he argued,

> The jobs here come "through relations." If I have a relationship—if you do, sometimes it's political, for the most part, because the following happens: Dos Rios needs workers, and the president of the *junta de vecinos* has a meeting with them, and they talk, and the company asks the *junta* to find workers . . . So the president of the *junta* comes and says, "I need this and that and you, you, and you go," and this happens each season. The governor does the same thing; when they need workers for Falconbridge or Barrick Gold,

the "relations" look for people, and they put out the word, or they send a list . . . and then they have meetings. And what happens in those meetings? In those meetings, if there are 20 people, at least 15 of them are people with connections to those in political offices.[24]

GRASSROOTS PERCEPTIONS OF RESPONSIBILITY FOR COMMUNITY WELL-BEING

In Bonao, just under half of those interviewed believed that government was the entity primarily responsible for community well-being (i.e., seven out of 16 respondents in Bonao, compared to just over half in Villa Altagracia). A homemaker and former free-trade-zone garment worker whose two children had received surgeries through Hanes's program argued that the "government and firms" are together responsible for community well-being,[25] but neither she nor anyone else I interviewed in Las Delicias answered that companies were solely responsible for community well-being.

The woman farmer I interviewed argued that "principally, public works in the community and other things—it's the government that has to deal with this, with governance, public works, public health, local government—and people have to demand these things because it's our right, as human beings and as communities, to have our necessities and those of the community fulfilled."[26] There was a common sentiment that individual people had a responsibility to make demands on government to deliver services and that they would be more effective in working together. As a stay-at-home mother explained, "people ourselves have to be engaged in making demands that government deliver on its promises and that it work on behalf of people."[27] A local schoolteacher whose child had also been operated on through Hanes's program argued more broadly that "everyone together" should work toward community well-being because "this way we benefit more."[28]

Yet there was a considerable level of skepticism about government's willingness or ability to help poor people facing unemployment or underemployment. As a security guard nearing retirement age observed, "there is not a lot of help in finding work—only if you are politically connected. If you want to work, you have to go through them, the politicians, because they control everything."[29] Only one person in Las Delicias, a secretary for a local nongovernmental organization (NGO), answered that people themselves should be principally responsible for their own well-being, in contrast to almost a quarter of the sample in Villa Altagracia who placed the responsibility for community well-being on the shoulders of individual people. The secretary reasoned: "I may pertain to a company or I may not, so the companies have to contribute to the society, but we ourselves as individual people should also contribute to making society better."[30] In her opinion, the only type of outright obligation a company had for community well-being was in relation to damage control and remediation in the case of industrial pollution: "The society is the medium in which companies develop, so they have to contribute to helping the society improve. If, for example, Falconbridge contaminates the environment, then supposedly they have to plant trees in order to improve the environment. At least in this way, they can improve something that they've damaged."[31]

Two people in Las Delicias argued that NGOs have a key responsibility for community well-being, whereas no one in Villa Altagracia looked to NGOs to take the lead. As a local childcare worker explained, NGOs were primarily responsible for community well-being because "here the one who in practical terms is most able is the *junta de vecinos*. This community is, like, isolated—and no one worries about Las Delicias."[32] The role of the *juntas de vecinos* was thus pivotal to perceptions of how work opportunities arise (or do not) in the neighborhood, since the *junta* has become a vehicle for distributing benefits—whether they come from the government or from private-sector entities.

Being at odds with leaders of a neighborhood association could mean blocked access to benefits or work opportunities. A mother and homemaker noted that for those with an education there are more opportunities, but qualified people are not always given the chance to apply for good jobs because the *junta* controls the process and the jobs often go to the "people with connections," not the most deserving.[33] As a laundry worker explained, "the company soliciting people, they come to the *junta*, which has been told, 'We need this many young people, see what you can do,' and she [the president of the *junta*] gets the young men of that age together ... And some have success in entering, at least they do because of this lack of qualified workers. Perhaps it's not the group that everyone would pick because not everyone feels good about everything, but that's how" it is.[34] Information about work "comes through the *junta* . . . they call the president of the *junta*, and they tell her to get people together because they're going to collect résumés, and they will look them through and decide which of the most appropriate they'll take—five from here, five from the other side of the highway, and that's how it goes," explained the childcare worker.[35]

While multiple respondents spoke of the *junta* reviewing the résumés of workers in order to select the most qualified to put forward to Hanes for interviews, one's relationship with the president of the *junta* was equally if not more important. A retirement-aged woman I interviewed referred to the president of the *junta* as its "owner," explaining, "The *junta de vecinos* receives a lot of young people—they take applications and review them. Look, this daughter of mine who works in a lottery stall asked the *dueña* [owner] of the *junta de vecinos* for help in getting a job in the zone, but they told her they couldn't. I don't know why."[36] Another older woman, reflecting on the construction of a public childcare center in the area, felt aggrieved by what she perceived as the *junta* president's favoritism in awarding jobs in that setting:

> There were a lot of women here at home who would've preferred to work there. We would have liked to—but then they chose those who they wanted. . . . They choose people because of personal relationships—not based on who wants to work. . . . It's the president of the *junta* who decides; she decides who to choose, and whether there's any more. . . . She gets things for her people, and even if we live close by here, it doesn't go to us.[37]

The owner of a small corner food shop (*colmado*) pointed out that there was one *junta* "above" the highway, in the main part of town, but he was vague about its functioning, noting, "I don't think they're getting together now."[38] The woman farmer pointed to the relative privilege of the *junta* on the other side of the highway: "When they get things up there, for the most part they do what they will and they don't tell anyone down here."³⁹ This division between neighborhood associations "above" and "below" the highway as well as among members (current and past) of the *junta de vecinos* in the neighborhood of Las Delicias was particularly pronounced—reflected not only in local discourse, but in government data as well.

Prior to my field visits to both Villa Altagracia and Bonao, I gathered quantitative data from a representative in charge of NGO relations for the Ministry of Economic Planning and Development (MEPyD), based in Santo Domingo. There are upwards of 7,000 NGOs nationwide, of which 1,500 are accredited to receive government funds. The majority of registered NGOs are based in the country's capital, Santo Domingo. MEPyD is currently compiling a national database of registered NGOs across all states nationwide and is working to build administrative capacity among them (with support from the World Bank). The full list of all registered NGOs in the state of Monseñor Nouel includes 165 groups, 66 of which are based in Bonao, while only two are located in Las Delicias (i.e., the orphanage and a local charitable group). All the rest of the registered NGOs in Bonao on the lists I obtained—from sports clubs to nursing homes and women's shelters—were clustered in neighborhoods on the other side of the highway.⁴⁰

The *junta de vecinos* in Las Delicias was not included on the MEPyD list. As a woman who had played a role in community organizing for some time explained: "When the *junta de vecinos* gets legalized, when it's an NGO that you can find on this list, because it's legally registered, then this *junta* will have the skills and right to make demands" for food, medicine, or clothing "because by law, it's an NGO and it has to act like one.... [It] could ask for a high school on this side of the road, because we don't have one, and then the kids wouldn't have to cross the highway. Go look for a ball court in this community, where there's a huge number of adolescents and youth—there isn't one, despite the fact that there's plenty of terrain and the government could simply say, 'Go and buy three hectares and make a court.' No, it's not here, because there aren't people with a vision for the future, who can say, 'Let's do this and that because we have to achieve something in this community.'"[41] Registered or not, the *junta* in Las Delicias nevertheless continues to play a significant role in organizing demands vis-à-vis local government officials and in acting as a liaison with Hanes management in the Dos Rios factory zone in the hiring of workers.

CONSTRAINTS ON STAKEHOLDER DIALOGUE

In Bonao as in Villa Altagracia, I structured my interviews with people in the community in an effort to gather comparable data on life and work histories along with their perceptions of responsibility for community well-being; availability of information on corporate practices in their community; familiarity with or willingness to take part in stakeholder dialogue; and attitudes on the nature of remedy. Residents of Las Delicias responded to questions about the availability of information on company activity by pointing to a similarly wide range of sources as did those in Villa Altagracia. In Las Delicias, roughly a third of the sample (i.e., five people) received information by word of mouth and another third (i.e., four people) through the local *junta de vecinos*. Three people relied on traditional media such as radio for their information on corporate activities locally. One person relied on Internet-based listservs; one had first-hand experience with Hanes by virtue of his employment there; and two had no idea how to obtain information about corporate activities. Among the most reliant on radio were older and/or homebound people. Two retirement-aged men and at least one woman with medical difficulties mentioned that company activities are covered in the media, particularly on the radio.[42] Two older women responded that they either did not watch the news or were not interested. One of these women relied on her husband for such information: "He knows more than I do because he watches the news," she explained.[43]

In contrast, a young mother who works as a secretary for a local nonprofit organization explained that she receives an electronic newsletter via cell phone that is full of information on jobs in both the public and private sector along with notices about training opportunities. Having attended a local technical school, she signed up for the newsletter while a student and continues receiving it. Like two young men I interviewed in Villa Altagracia, this young woman was tech savvy and well networked socially. As she explained: "Most of the information comes from networks this way; in fact, I just got a notice for government jobs in the judicial branch—they posted for various vacancies" she said, showing me the newsletter link on her mobile phone. "I joined this group listserv and the announcements come to me that way," she explained. "It's a WhatApp group. And so, if I get information I send it to the group, and the one who's interested follows up. They set up the group when we were in our university career, and we sustain it."[44]

Like her counterparts in Villa Altagracia who participated in listservs to gain information about companies (particularly in relation to hiring), this woman was in her late 20s. She had a higher level of education than most other people I interviewed in Las Delicias—in her case, the equivalent of an associate's degree in business management. In the case of Internet users whom I interviewed in Villa Altagracia, two had skilled trade certification through INFOTEP[45] and three were part-time college students or preparing to enter college,[46] while another was a union truck driver.[47] In both towns, those with social media skills sought to use them to increase their employment opportunities (or in the case of independent artisans, to market their products or services). As the secretary in Las Delicias explained: "In my area, you can find work, so that's how it is. . . . A cousin of mine who also studied administration has worked for several years in the free-trade zone, and from there they transferred her to Santo Domingo, to the headquarters of the company. . . . Now she's the head of human resources. She studied administration and then did a master's degree in human resources."[48]

Despite individual efforts to obtain information on corporate job opportunities, many people remarked on the difficulty of doing so in a context where corruption was pervasive. The perception that corruption stymies development and constrains personal opportunities was a consistent theme across both towns. Notably, my fieldwork in the Dominican Republic in the early summer of 2017 coincided with the growth of an island-wide social movement focused on government corruption known as the *Movimiento Marcha Verde* (Green March Movement). Protesters have organized simultaneous marches in multiple towns across the island, pressing for political reform, government transparency, and an end to impunity. At least one of my respondents (a well-known feminist unionist in her 60s, whom I interviewed in Santo Domingo) had taken part in marches.[49]

Getting information on jobs is still principally linked to currying favor with people in power, no matter what the level—and this, argued one of the men I interviewed, leads to financial corruption. If information on jobs were shared "through fliers or public announcements and this type of thing, it'd be beautiful, because if I saw something in the newspaper, I'd go" to interview, he asserted. But in his community, information on company activities moves "mostly mouth to mouth, but for it to go mouth to mouth you have to have it move pocket to pocket . . . it's through connections, as I referred to it, through deep pockets."[50]

ASSESSING LOCAL KNOWLEDGE OF AND RECEPTIVITY TO STAKEHOLDER DIALOGUE

In addition to lack of information and pervasive corruption, there were other factors that influenced respondents' knowledge and receptivity to stakeholder dialogue in Las Delicias—with interesting similarities to and distinctions from those in Villa Altagracia. Integral to my questionnaire and the broader cross-case comparative framework, I used three questions to gauge awareness of and receptivity toward stakeholder practices, including a factually based question (whether the person or anyone they knew had ever met with company representatives); a hypothetical question (whether the person would be willing take part in stakeholder dialogue); and a conditional question (on the type of resources the person would need and the priority-setting she or he would do in order to make the interaction effective).

No one interviewed in Bonao had directly participated in stakeholder consultation, despite the presence of many more types of industry than in Villa Altagracia. Aside from awareness of protests related to environmental issues in mining, there was a relatively low level of awareness of company practices related to stakeholder engagement. As a security guard of retirement age explained: "It's like I told you: you don't get a lot from companies. The people who are high up there only help themselves, not others." If invited to participate in the dialogue, he viewed it as an opportunity to bargain for his own workplace benefits—a personal negotiation, rather than a vehicle for remedy at the community level.[51] His wife did not understand questions related to stakeholder dialogue, nor did several of the other women I interviewed.

Some people in Bonao conflated the idea of stakeholder dialogue directly with the work of the *junta de vecinos*. As a homebound woman with significant health problems explained, "I've never heard of this. No one [from the *junta*] has visited me."[52] Others thought that dialogue with a company could only happen in the context of a work relationship, as did the retirement-age security guard, who said he would bargain for vacation benefits if involved in such a consultation.[53] Others believed that stakeholder dialogue was somehow a political act or was connected to formal political parties. A *colmado* owner said he would be open to participating "because the majority of people here do things casually to try to get something out of it, at least those participating in the *junta de vecinos*." But he also noted the potentially political nature of the process: "if they get together the people in the neighborhood, it may also be for political meetings."[54]

Among the people who expressed interest in taking part in stakeholder dialogue were a man involved in finance work and a schoolteacher, both of whom appeared to have a higher SSS linked to the comparably higher status of their occupations and correspondingly better homes.[55] A laundry worker in Las Delicias, whose husband is employed in the telecommunications industry and whose two sons have worked for or are currently employed by Hanes, was also willing to engage in stakeholder dialogue, but her willingness to do so was conditional on what the theme of the talks would be.[56]

For those who understood what stakeholder dialogue was and were at least in principle willing to take part in it, the common perception was that more information would be needed in order to make the process effective. Two-thirds of those interviewed in Las Delicias emphasized the importance of understanding community needs before taking part in dialogue and would justify their participation as part of meeting community needs. A public school teacher explained that "by participating in this type of meeting, I could help find a solution to community problems happening now."[57] The laundry worker explained that she would "need more information on the theme of the consultation" before being willing to participate—in part, to be able to get the most out of the process and in part to avoid wasting her time.[58]

Despite Hanes's efforts at promoting its medical philanthropy in Bonao, there was skepticism in the neighborhood of Las Delicias about the quality of the Marchena Hospital, voiced by several people I interviewed. As an older homemaker remarked, "This hospital doesn't work for anything.... No one gives it anything. They bought a new parcel of land nearby and they still haven't used it for anything." When I asked if she had heard of the Hanes-funded surgeries, she replied, laughing, that there's "nothing like that—I don't hear about it, I haven't heard about it, and I won't hear about it in the future!"[59]

In a similar manner, a young mother and homemaker laid out the main priorities that she would want to articulate if she were involved in a company–community dialogue. First, she would stress that the problem of electricity shortages is chronic and needs to be addressed. Second, "we need a better hospital—the Marchena is pitiful!" (By way of justification, she argued that the Marchena is so outdated and poorly maintained that the only reason "that they have even started to create a site for a new hospital is because the community protested and had a communal strike. The minute things settled down, the hospital stopped moving forward with construction." In her estimation, the only reason anyone goes to Marchena is for emergency care; everything complicated has to be handled in hospitals in Santiago or La Vega. For routine care, people go to private clinics in Bonao.) Her third priority would be to call for basic facilities "on our side of the highway," like a grocery store, an urgent care health center, or a high school. Finally, she argued, the community of Las Delicias needs a highway overpass so that people can safely cross the road to the rest of Bonao. If she were to be involved in a stakeholder dialogue, this is one of the main priorities she would signal.[60]

COMMUNITY ATTITUDES ON THE NATURE OF REMEDY

Key elements from the list of priorities outlined by this young mother were echoed across multiple interviews: health care, electricity, transportation, and crime were all common concerns. But connecting them in any way to a right to remedy was a step most respondents did not make, either in Bonao or in Villa Altagracia. In both settings, I included a short list of statements concerning the nature of remedy. I asked respondents to tell me whether they agreed or disagreed with them, and why. One statement was that companies do not have social obligations beyond their employees or shareholders; another was that communities should accept whatever contributions companies make and should not expect more; a third was that labor unions can only help their members, not other people beyond them.

Far more people in Las Delicias than in Villa Altagracia responded that companies had a social obligation to the people in their communities beyond their employees, by a margin of nine to one (with another five others answering they did not know and one not answering). As the woman who worked as a secretary in a local NGO argued, companies "have a social responsibility. They should return to the community something of that which they take from the community. If a company develops within a community, it can also help this community better itself."[61] As a mother and homemaker similarly argued, companies "have to give back to the place where they make their profits."[62]

Some respondents in Las Delicias believed that particularly when a company damages the environment or in some other way causes harm, then it is obliged to help. As the laundry worker argued, companies "should give for the necessities in the communities. It's like I told you: Falconbridge and Barrick are huge companies, huge, and they make many millions of pesos and they should pay for the nature that they're exploiting. But at the same time, it doesn't matter to them—they ought to do this, but they don't go around saying, 'there's a community which needs such-and-such a thing, like a basketball court.' At least they should give this, because they've contaminated the nature around them."[63] But she, like some others in the community, believed that in routine circumstances, if the company gave anything at all, the community should be satisfied. Across all interviews in Bonao, Hanes was never placed in the same category with extractive companies, thus leaving any discussion of remedy Hanes might owe the community off the putative negotiating table.

People I interviewed in Las Delicias were evenly divided among those who believed that if the company made a contribution to the community, the community should simply accept it and not demand more, and those who did not (i.e., five agreed with that statement, five disagreed, and the rest responded that they didn't know how to answer). The Hanes worker interviewed near his home argued, "if there's no urgency, then you accept it. If there's a donation, you can't demand more."[64] He was particularly quick to note Hanes's efforts to address environmental impact—not surprising in a community where environmental damage related to the mining industry has been a significant issue.[65]

The secretary argued, however, that a company should "keep helping" beyond an initial donation, because with a single donation "they don't solve everything."[66] Other people, however, argued that because Hanes is a private company, it has a moral but not a legal obligation to help anyone other than its workers. As the female farmer and grassroots organizer argued (referring to Hanes in shorthand form by the name of the export processing zone where the factory is located):

> Zona Dos Rios is a private company. Now, people have asked the Zona Dos Rios for lighting and other things and they've cooperated in a major way. And when the school year begins, Zona Dos Rios gives backpacks, notebooks, pencils, and lots of things. But it's a private company. Now who has the responsibility? Principally, the state of Monseñor Nouel . . . For me as a human being, if there's someone who has a company and [he or she] decides to help, seeing the situation, this comes from the person themselves. *I think that as a community, we shouldn't oblige the company to help. If there are conscientious people in the company, one can go to them . . . But this they do voluntarily, not an obligatory sense.*[67]

For some of those I interviewed, the risk of accepting donations of any sort was that it could lead to cooptation by corrupt actors. As the man who worked in finance cautioned, "if someone is bribing me to keep my mouth shut, and I don't say anything—that's what's happening these days. It's a bribe to take it and keep quiet. If they give me something and ask for something in return—don't give me anything! Give me what corresponds to me alone, or they'll keep bothering me forever."[68]

The sample was similarly split (i.e., five to five, with the remaining respondents answering "don't know") concerning whether unions had any responsibility to people beyond their members—in this case, more closely mirroring attitudes in Villa Altagracia. As an older man who had worked for years driving a motorcycle taxi argued, "Unions don't help anyone. No one! The most they do is plead the case of the worker if he gets sacked from a company. The lawyers for the union will make sure that the worker receives his severance and bonus."[69] The owner of a *colmado* was more circumspect, explaining: "Here there aren't unions—in the *zona*, yes, all the companies form unions. But the majority worry about their members. But if I don't belong to a union, then it doesn't have responsibility for me."[70] By contrast, a childcare worker argued that unions should look beyond their membership to support a broader range of people in the community: "They can't just worry about their own employees, of their own company—what about the rest of us who don't have a union? I don't have one, so who worries about me? I worry about myself. I'm my own union."[71]

These mixed views on unions were emblematic of variation in attitudes among people in Las Delicias; this small, tightly knit neighborhood on the "other side of the highway" was one with considerable differences among individual people in terms of both SSS and life experiences. Yet there was an overall tendency among the people I interviewed in Las Delicias to expect government, NGOs, companies, and unions to all have a responsibility for doing no harm—even if this was not connected explicitly to a sense of entitlement to remedy. There was also considerable attention to the role of the *junta de vecinos* as a mediator among government, companies, and the community, which was more pronounced in Bonao than Villa Altagracia. Untangling the factors that have contributed to these tendencies is central to understanding the potential and limits of existing models of stakeholder consultation more generally.

LESSONS LEARNED

In a vertically integrated production system, where textiles move from plant to plant, "collar to band,"[72] *Hanes's decision not to acquire the Alta Gracia company means that the idea of a "living wage" has not circulated from plant to plant—even with such a high-profile example less than an hour away.* Workers from Hanes are highly visible in a small neighborhood like Las Delicias, and people I interviewed were more overtly desirous of jobs at Hanes in part because there were so many more potentially available here than in Villa Altagracia (i.e., 3,000 as opposed to 200).

The role of the junta de vecinos *as gatekeeper was more visible because it acted as an employment broker, mediating interaction between the company and community.* This complicated the nature of SSS in Las Delicias: being in the good graces of the *junta* was integral to having access to work and, with it, to further INFOTEP training paid for by Hanes. While Adler-Milstein and Kline have noted the higher SSS of Alta Gracia workers relative to their neighbors (2017, 110, 118, 122), in Las Delicias resentment was more often directed at the *junta* than at companies themselves, and respondents to some extent judge their own fates in relation to the *junta* rather than by the presence or absence of work or benefits offered by Hanes.

The corporate philanthropy that Hanes carries out in the medical sector is not overly visible; no one I interviewed made a consequentialist link between labor unrest and Hanes's subsequent engagement in corporate philanthropy. Instead, the "low bar" for corporate social responsibility (particularly in an area where extractive industries have had a troubled history of environmental contamination) has meant that simply gaining access to work was viewed by many respondents as a short-term goal and the level of expectation of companies was largely to do no harm, rather than to be responsible for remedy per se.

Many of the people I interviewed in the neighborhood of Las Delicias had internalized the process of needs assessment and priority-setting. When asked hypothetically about stakeholder dialogue or remedy for human rights problems caused by companies, these respondents were quick to list priorities for action. A baseline community assessment (a needs analysis) is standard practice in much development work, and the neighborhood of Las Delicias had already been host to several generations of Peace Corps workers, each of whom conducted their own community diagnostic survey before commencing their formal work in the community. The childcare worker I interviewed argued that if she took part in a stakeholder consultation, she would want to be able "to demonstrate the needs of the community. The first priority should be the children and young people. . . . We don't want young people who have vices, young people hanging out . . . Here, if you don't work you can't study, because the tuition may be free, but you have to buy other things."[73] Others cited the troika of jobs, infrastructure (both electricity and a footbridge over the highway), and crime as their primary concerns.

Community members often viewed politics as a zero-sum game in which those with more resources either overtly or covertly (through corruption) sought to maximize their own gains at the expense of others. Stakeholder dialogue in a context like Las Delicias thus became associated by respondents with an opportunity to assess how much support a company might be in the position to afford the community—and to bargain. As one housewife explained, she would take part in a dialogue "because I would want to promote community well-being. You won't know what's there if you don't knock on the door."[74]

Finally, gender disparities persist and in some ways were more pronounced in Bonao than in Villa Altagracia. The women workers I met from Alta Gracia Apparel had played leadership roles, whereas women in Las Delicias commented on the lack of professional options and the double burden of work in the home and paid sphere as well as the relative lack of opportunities for women in their neighborhood. In both towns, moreover, INFOTEP courses in skilled trades were typically the prerogative of men, not women.

IMPLICATIONS

Observations from both of these communities reveal just how complex the context is for contemporary stakeholder dialogue. In Las Delicias as in Villa Altagracia, government faces a significant credibility gap at the grassroots level; engaging government as a partner in stakeholder dialogue and in delivering remedy jointly with companies is key to the sustainability of interventions in the health sector or in relation to infrastructure. Moreover, the disdain for government is understandable in a context where corruption is pervasive, as the social mobilization around the *Movimiento Marcha Verde* demonstrates and as the Dominican Republic's poor score in global corruption rankings demonstrates—135th out of 180 countries ranked (in the bottom quarter, among the most corrupt globally), as rated by Transparency International (2017).

People in Las Delicias had a less expansive vision of alternatives to "business as usual" than those in Villa Altagracia—in part because Hanes is operating at full force as a source of conventional factory employment in Bonao, thus reinforcing a conventional view of workplace governance and corporate remedy for economic rights shortfalls. By contrast, in Villa Altagracia the crisis of unemployment and underemployment has led to worker-driven alternative forms of employment like the Alta Gracia factory and to even more radical rethinking of employment beyond existing business frameworks (such as the vision shared by the artistic entrepreneur in Chapter 4).

Neither of these communities is homogenous, but both face the common question of whether it is possible to reconcile the two faces of BHR—namely, the shop floor/workforce face versus the one that community members see. Placing BHR efforts within the context of remedy for broader underdevelopment and underemployment—particularly in towns where light manufacturing takes place—is an ongoing challenge that I grapple with in the next chapter, on policy alternatives to business as usual.

CHAPTER | 6

Policy Implications of Changes in Stakeholder Consultation

IN TOWNS LIKE VILLA ALTAGRACIA and Bonao and in thousands of other communities like them where global supply chains extend, the challenges of sorting out obligations for economic rights fulfillment and ensuring effective remedy persist. While workers may be in a position to make claims on employers, average community members cannot easily do so. Often, people who do not work directly for global firms or their suppliers are unable (or unwilling) to make rights claims on the companies in their communities, as revealed in interviews central to Chapters 4 and 5 with residents of Villa Altagracia and Bonao, respectively. Yet many of those same community members have a strong sense of their citizenship rights, as conveyed in interviews where they often voiced frustration over limited political will and/or state capacity to address the structural roots of poverty in the context of global manufacturing.

Community members do not figure centrally within most conventional theoretical or policy frameworks for stakeholder dialogue. Because stakeholder mechanisms remain concentrated in the extractive industries instead of light manufacturing (as discussed in Chapter 3), many of the remedies developed do not address the structural factors underlying poverty in manufacturing communities—such as downward pressure on wages, persistent constraints on unionization in light industry, and consumer and brand pressure for lower prices.[1] The third-party auditing and human rights compliance programs central to contemporary supply chain management in light manufacturing tend to reinforce these blind spots (Anner 2017, 64; LeBaron, Lister, and Dauvergne 2017), touching tangentially, at best, on community engagement. In the process, they often miss the depth of conflict over economic rights within the communities where workers live alongside other people affected directly or indirectly by corporate practices ranging from environmental management to hiring protocols to decisions to relocate or close a facility.

This chapter takes community members' position in stakeholder dialogue as its point of departure. It analyzes evolving policy approaches from the standpoint of a broader range of members of the community. It explores mechanisms for enforcing shared private-sector *and* state responsibility for structural remedies to address poverty in light manufacturing communities. Given the comparable lack of engagement of grassroots voices in the design and implementation of remedy in the business and human rights (BHR) arena, as discussed in Chapters 2 and 3, this chapter assesses both the potential and limits of contemporary policies in terms of how they address the pervasive inequality among people involved in and affected by light manufacturing activity globally.

Specifically, the chapter takes stock of two distinct policy approaches to brokering interaction among companies, workers, and communities in the context of global supply chains, namely multi-stakeholder initiatives (MSIs) and an alternative known as "worker-driven social responsibility" (WSR) models. The former are the prevailing corporate-led approach and the latter have emerged as a result of grassroots and popular mobilization. I analyze the two emerging pathways (i.e., MSIs and WSR models) from the standpoint of community engagement and impact. Other scholars have explored sources of variation in how public- and private-led governance models affect protection of labor rights and social standards in the context of the workplace (Fransen and Burgoon 2017; Marx, Wouters, Rayp, and Beke 2015), but this chapter explores whether differing approaches to stakeholder engagement have the potential to affect the degree of leverage that community members have in negotiating with companies and with states.

Rather than juxtaposing MSIs and WSR models as stark alternatives, I explore potential hybrid approaches to creating consultative processes that give community members a greater say over the terms of who participates, what the rules entail, how information flows to parties involved in dialogues, who monitors outcomes, and what constitutes abuse and remedy.[2] This chapter thus draws not only on case study findings from the Dominican Republic but also on participant observation carried out in the context of an international conference hosted by the University of Connecticut Business and Human Rights Initiative in October 2017, where academics, advocacy and labor leaders, and business representatives from across the

United States, Europe, Asia, and Latin America analyzed the emergence of these two models and the challenges inherent to each.[3]

TWO PATHWAYS, MULTIPLE CHALLENGES

Grounded in the Ruggie Principles' emphasis on remedy and organized around conventional MSIs, the first approach is integral to many existing corporate social and environmental compliance programs. The United Nations Conference on Trade and Development's World Investment Report 2011 (2011, 112) characterizes MSIs as "cross-sectoral partnerships created with a rule-setting purpose, to design and steward standards for the regulation of market and non-market actors."[4] Legal scholar Jennifer Gordon explains that most MSIs "bring together businesses (especially international companies with globally recognized brand names), nonprofit organizations, and sometimes governments, to set new voluntary standards in problem areas that have proven difficult for national governments to address effectively" (2017, 2). Some MSIs attempt to broker resolution of community grievances, while others serve as a vehicle for managing interaction between companies and local groups involved in philanthropically oriented community development. This is the nature of the policy landscape in Bonao.

The WSR approach centers on workers themselves driving the creation, monitoring, and enforcement of rigorous, legally binding standards for supply chain management. To do so, they collaborate with a different set of nongovernment actors, including staff of worker centers (Fine and Gordon 2010; Nolan 2018), unions, and community-based organizations, in order to create systems for enforcing the social and environmental sustainability of global supply chains. Workers carry out extensive peer education, monitor for violations, and file complaints that trigger enforcement. Employers at the top of the supply chain are mandated to "pay subcontractors for the improvements that code compliance requires" to eliminate the risk of brands either offloading responsibility to players further down the chain or exiting when crisis arises (Gordon 2017, 7).[5] The Alta Gracia factory emerged as a result of this second type of approach to corporate responsibility, brokered through the Workers' Rights Consortium in conjunction with local Dominican unions (Adler-Milstein and Kline 2017).

This chapter compares these two modes of stakeholder engagement (MSIs and WSR), takes stock of the challenges facing each, and explores potential synergies. A growing number of scholars, policy practitioners, and human rights advocates agree that corporate–community consultation "processes should be designed and approved by the affected persons—the rights-holders—rather than those who are believed to have caused the problems in the first place" (Kaufman and McDonnell 2015, 128; see also Wilson and Blackmore 2013). In turn, some analysts and advocates focus on reforming existing MSIs to be more transparent and inclusive. They recommend creating community-led impact assessments, community-designed impact and benefit agreements, and community-driven free, prior and informed consent procedures in order to protect and empower the people most vulnerable to human rights violations in the context of international business (Kaufman and McDonnell 2015, 129–130).

Public law scholar and advocate César Rodríguez-Garavito finds that MSIs "tend to both encourage the engagement and limit the participation of relevant actors," in part because consulting firms play a key role in "amplifying the voice of more powerful actors (e.g., corporations and states)" disproportionately over those of "less powerful stakeholders, such as affected individuals and communities. And the chances for accountability politics to serve as a source of upward pressure [on norms] in new governance regimes may be consequently reduced" (2017a, 31–32).[6] This distinction between categories of civil society–based groups (professional auditing/research organizations vs. grassroots groups) is echoed in the broader literature on the challenges related to grassroots participation in stakeholder dialogue.

Critics of conventional MSIs, such as business scholars Prakash Sethi and Janet Rovenpor, argue that key nongovernmental organizations (NGOs) active in the social auditing sector are vulnerable to "managerial capture" (2016, 6) by the businesses they interact with—which, in turn, impedes these MSIs' ability to mediate on behalf of workers (and, by extension, community members).[7] Legal scholar Tara Melish has argued more broadly that the majority of post-Ruggie mechanisms and processes are inherently constrained by design: "The GP's [Guiding Principles on Business and Human Rights, or Ruggie Principles] limited 'remedy' prong provides little meaningful relief in this context. It does not entitle affected communities to engage in problem-solving, agenda-setting, and regular monitoring of business-related human rights harms, but rather only to seek (allowable) redress once discretionary abuse has already occurred" (2017, 84).

WSR approaches have thus emerged as an alternative to MSIs. Proponents argue that WSR models have the potential to shift power and to expand the scope and effectiveness of remedy. They can counter the pervasive lack of specificity[8] over what constitutes robust and inclusive standard-setting and enforcement of labor rights as well as what constitutes meaningful consultation. WSR models are explicitly designed to engage workers in setting the baselines for codes and monitoring practices. They pivot around the participation of workers in peer-to-peer training and ongoing needs assessments that embed corporate responsibility practices in local, social practice. This view of supply chain governance—not as being imposed from above by international brands but as being co-constructed from below by workers and community members in dialogue with companies—marks a distinct shift from prevailing practice. It reflects a broader movement in the human rights field toward deeper participation and engagement of local-level actors, norms, institutions, and strategies that can transform the meaning, goals, and outcomes of human rights practice writ large.

Critics such as Samuel Moyn (2010) and Stephen Hopgood (2013) have argued that human rights laws and corresponding state and nongovernmental institutions and actors are out of touch with (or out of reach for) most marginalized people and are thus ineffective at reducing global inequality or confronting violent forms of social exclusion and conflict. By extension, one could argue that stakeholder dialogue practices exhibit all these flaws. But economic and labor rights promotion through WSR models and other forms of participatory rights practice challenge both Moyn and Hopgood's critique (Albisa 2018; Saiz 2018). For example, Brooke Ackerly has forged a theory of "just responsibility" for human rights (informed by her longstanding relationships with garment workers and their advocacy allies) that hinges on individual people taking political responsibility for global injustice through just actions of many kinds, including but not limited to ethical consumption. The theory speaks "to the student and the professor of Peace Studies, to the parent outfitting a child for school and to the laborer making the school clothes, to the garment labor rights activist and to the global retail shareholder, to the domestic worker and to her employer. To each a theory of just responsibility offers an avenue for political engagement, not merely an opportunity for ethical reflection" (Ackerly 2018, 16).

Similarly, Michael Goodhart argues that human rights advocacy rooted in a "praxis of egalitarian freedom" (2018) has the potential to empower people to "enact, articulate, and even transform their own (and others') understandings of themselves as citizens and as members of communities" (2018, 8 of proofs). It can also generate a "distinctive form of power by fostering a sense of agency and self-esteem among participants as rights-bearing subjects" (2018, 8). Both Ackerly and Goodhart are committed to building human rights theories useful in achieving social transformation—including and especially within the domains of economic rights and justice. But as Ackerly acknowledges, "The methodological challenge is: how do we explore the complexities of oppressive structures—revealing them and deconstructing their pernicious functions—without becoming mired in the process of deconstruction *at the expense of addressing injustice itself?*" (2018, 39; emphasis added).

This chapter is similarly motivated by a pragmatic desire to find workable solutions to human rights violations in supply chains by involving a wider set of actors in constructing just solutions. MSIs face constraints on popular participation, but WSR models also face challenges—specifically, those of replicability and scalability. The fast-paced, highly mobile nature of contemporary manufacturing is at odds with the deliberative and consultative nature of WSR models. Prevailing wage rates are far out of line with the egalitarian goals of WSR. The scope conditions for successful WSR include the existence of strong community-based organization with the "capacity to engage" (Fine 2017, 364) government, business, and social movement allies. But NGOs are often lacking in number and capacity in many communities where light manufacturing takes place. (Chapters 4 and 5 illustrated the dearth of local NGOs and the internal divisions among neighborhood associations that complicate local organizing in Villa Altagracia and Bonao.) Policies aimed at making stakeholder dialogue more inclusive and just (in Ackerly's terms) would thus seem inherently constrained.

Yet innovation has occurred even in the most seemingly intractable settings—and in ways that straddle the MSI and WSR divide, as one prominent example far afield from the Dominican Republic demonstrates. In Bangladesh, the audit and compliance model for ensuring factory safety in the garment industry in that country—the Accord on Fire and Building Safety in Bangladesh—emerged in the wake of the world's worst industrial disaster in the garment sector in 2013 (Reinecke and Donaghey

2015, 257–277). Following a factory complex collapse that killed an estimated 1,100 workers from multiple factories and injured more than 2,000 others, the brokering of this tripartite agreement has resulted in legal guarantees that the cost of remedy for factory safety be borne by brands, not be passed on to workers in the form of lower wages. The Accord has involved over 2 million Bangladeshi workers centrally in peer-to-peer education and onsite implementation of ensuing factory safety measures (Accord on Fire and Building Safety in Bangladesh n.d.). Ackerly's work traces the involvement of locally-based NGOs and their transnational allies in bringing the Accord into being over multiple years (Ackerly 2018), and key NGOs in the WSR movement claim it as a success—including the Workers' Rights Consortium, International Labor Rights Forum, and others. But businesses with standard third-party auditing programs in the MSI vein *also* point to the Accord as a vehicle for more effectively managing supply chains.

Signing the Accord is now integral to many conventional supply chain management programs run by brands that do business in Bangladesh. It is a condition of doing business with many universities involved in producing collegiate apparel. Many participants at the October 2017 University of Connecticut Business and Human Rights Initiative conference referenced the Accord as a successful example regardless of which side of the aisle (MSI or WSR) they sat on.[9] As the policy paper produced following the conference notes:

> Participants highlighted the quality of scholarly and policy analyses of challenges in Bangladesh's textile industry, which illustrate why long-term strategic partnerships are necessary, given the forced tradeoff between productivity and

human rights endemic to "fast fashion." Re-negotiating the terms of timeliness and creating purchasing practices which enhance trust are central to crafting alternative business models. (Leipziger 2018, 18)

Donaghey and Reinecke emphasize the Accord's central requirements for the democratic participation of workers; for holding authority (corporate actors) to account; for due process in dispute resolution; and for "a balance of power between the employer and workers through collective organization" (2018, 16)—all of which are aligned with the principles of WSR but which are now moving into the more mainstream MSI arena through the Accord's demonstrated success at paving the way for significant improvements in factory safety and worker capacity-building. How the Accord will affect people *beyond* those employed in the country's garment factories is an open question. There is some evidence of carry-on effects, such as increased demand for higher-skilled/safety-conscious workers and a willingness of auditors to address gender-based workplace violence in the context of broader efforts at safety promotion (Bride 2018). Marshalling sufficient political will on the part of the Bangladeshi government to sustain the agreement's achievements remains challenging—including improving the transparency and efficacy of government inspections, enhancing public infrastructure, and expanding training and educational opportunities for workers.

Anner, Bair, and Blasi highlight governance parallels between the Bangladesh Accord and the Alta Gracia factory model, but these authors also signal the vulnerability of the Alta Gracia factory owing to the company's relative isolation from the rest of the garment sector in the Dominican

Republic (2013, 32–33). In Bangladesh, more than 1,600 factories monitored under the terms of the Accord are obligated by contract to uphold the same standards for fire and building safety and related worker training (Accord on Fire and Factory Safety in Bangladesh n.d.). By contrast, in the Dominican Republic, Alta Gracia Apparel stands alone in paying a living wage and thus faces competitive pressure from non–living wage factories domestically and abroad (Sethi and Rovenpor 2016, 15).

Bridging the MSI and WSR models may seem an insurmountable challenge in the face of the race to the bottom in low-wage, low-skilled, highly mobile manufacturing, where collective action is difficult to achieve among workers and even more so among other people in the community. Since the 1990s, however, there has been a parallel move toward creating legally binding guarantees of inclusion and participation of stakeholders in international treaty formation, driven often by people from marginalized communities themselves. Melish (2017, 89) highlights the success of people with disabilities and other groups at codifying the human right to participation in new treaty formation, implementation, and monitoring, including through the United Nations Convention on the Rights of Persons with Disabilities (2007); the United Nations Convention Against Corruption (2003); and the United Nations Economic Commission for Europe's Convention on Access to Information, Public Participation in Decision-Making and Access to Justice in Environmental Matters (2001). The codification of the right to participation could add normative heft to the WSR model while at the same time pushing the boundaries of conventional MSIs in new directions.

MAKING GOVERNMENT'S ROLE IN REMEDY CLEARER

The right to participate in stakeholder consultation could thus be considered an instrumental right that makes other rights (like the right to work, or the right to housing, or physical integrity rights) possible to achieve. If we accept this framing, then government also has a role to play in translating rights into practice in global supply chains—and doing so in line with human rights principles means placing special priority on the rights of the most marginalized people in any given community.[10]

As the case studies central to Chapters 4 and 5 reveal, people in communities affected by light manufacturing may have a strong sense of their citizenship rights, even if those citizenship rights are often made apparent more by their absence. In both Villa Altagracia and Bonao, roughly half of those interviewed in each community believed that government was responsible for community well-being—that is, just under half of all those interviewed in Bonao (i.e., seven out of 16 respondents) and just over half in Villa Altagracia (i.e., 15 out of 27). Indeed, the main connection between the two communities was the strength of citizenship-based claim-making on government.

However, workers and community members alike find it difficult to make claims on state actors whose interests often seem to them more aligned with those of companies than those of their own citizens. The lack of political will to remedy problems at the heart of stakeholder grievances and the lack of state capacity to do so are common features of places where light manufacturing supply chains extend globally.[11] The primary concern of many people in manufacturing communities like Villa Altagracia and Bonao is often making ends meet on a daily basis while grappling with challenges such as poor infrastructure and high crime. These types of problems affect workers and community members alike, though the workplace-centered remedies brokered by companies in the context of stakeholder dialogue are not often structural in nature (i.e., do not address economic rights violations beyond workers) or applicable to both groups.

Because workers have an employment relationship, they are entitled at least in theory to make claims in an organized manner on employers; by contrast, community members have less formal claim-making standing, unless they are able to organize to claim joint damages of some sort (such as in the case of environmental harm). Yet community members interviewed for this book often framed their primary concerns in terms of poverty, lack of employment, and crime; their subjective socioeconomic status was derived in relation to other people in the communities, including workers. Such constraints can and do limit the ability or willingness of nonworkers to participate in stakeholder dialogue with companies. Community members interviewed in both Villa Altagracia and Bonao were often unaware of the nature and scope of stakeholder dialogue practices regardless of the nature of the approach (WSR or MSI) under way in a given community. People who lack an employment relationship with a company frequently perceive themselves as unable to make callable claims on firms for which they do not work—other than remediation in the case of environmental damages.

The Alta Gracia business model thus reveals both the possibility and the challenge of expanding the form and content of remedy for structural poverty through WSR strategies. Even in a community where WSR resulted in the creation of a "living wage factory," people face the ongoing challenge of finding work that is paid a living wage *beyond* that factory itself, which employs just 200 people. The Alta Gracia business model appears better known outside the country than domestically, even just an hour up the highway in Bonao as well as in the capital of Santo Domingo. The challenge is to extend the living wage business model beyond the Alta Gracia factory (i.e., replicability) and to expand the model beyond the collegiate apparel sector to the larger textile market (i.e., scalability).

In Bonao, where the conventional MSI-cum-corporate-philanthropy approach prevails, the sheer scale of potential employment at the Hanes Dos Rios plant (with 3,000 workers) has meant that these jobs are still widely sought after by members of the community—even at the prevailing national minimum wage. (The company's medical philanthropy involving roughly 50 children annually touches far fewer people in the community directly.) Hanes's ongoing support for worker training programs administered through the Ministry of Labor's Instituto Nacional de Formación Técnico Profesional (INFOTEP) further increases the social desirability of these jobs. As discussed in Chapter 5, such training increases the competitiveness of the factory and others like it that are involved with INFOTEP across the light manufacturing sector. But these training programs are designed and implemented by free-trade-zone employers' associations (Schrank 2013, 306) not by workers themselves, with the consequent emphasis on enhancing productivity that leaves prevailing wage structures in place.

In my interviews with people beyond the edges of most consultative practices—community members affected indirectly by the factories in their communities in the Dominican Republic—it became clear that new strategies are necessary for engaging government as a party responsible for remedy, regardless of which pathway (MSI or WSR) companies take. Participant observation at the 2017 University of Connecticut conference revealed a similar emphasis on the need to re-envision the role for government in BHR frameworks (Leipziger 2018, 14) not only in relation to protecting fair wages but also in avoiding a regulatory race to the bottom (Leipziger 2018, 15). "The legitimacy and effectiveness of stakeholder dialogue hinge on its inclusivity, yet it remains a process that often involves only segments of those affected by or engaged in business activity (e.g., civil society *or* government *or* unions *or* business—but not all of these actors together, nor fully representative segments of each") (Leipziger 2018, 18; emphasis added).

Government representatives who participate in MSIs, in particular, "tend to be those of industrialized countries where many corporations are headquartered—rather than the governments of developing countries where manufacturing and sourcing actually take place" (Leipziger 2018, 6). The implicit assumption undergirding restricted participation is that innovative responses to the challenges of stakeholder engagement will come from the global North or from professionalized NGOs and research organizations. But involving representatives of a broader range of governments and community-based organizations in setting the terms of stakeholder dialogue widens the scope of potential solutions. More effective and inclusive stakeholder dialogue would thus entail moving the process beyond a harms-based approach toward a proactive one in which the right to participation becomes a constitutive element of remedy—where "employers, governments, international financial institutions, and *community members themselves*" collectively work toward "addressing the root causes of problems and searching for structural remedy" (Leipziger 2018, 7; emphasis added; see also Fine 2017; Albisa 2018).

BRINGING POWER-BASED ANALYSIS BACK IN

A key distinction between the MSI and WSR approaches is the extent to which they take as given the existing power dynamics that underlie global supply chains. MSIs have tended to take economic and social hierarchies as given, working to *reform* existing regulatory and enforcement systems to make these systems more transparent and fair. WSR approaches *contest* the basis of existing power relationships, seeking new forms of governance that expand the participation of people at the bottom of the chain and change the incentive structures of actors with power at the top.[12]

Rodríguez-Garavito (2017a, 24) has argued that the governance model the Ruggie Principles undergird (i.e., the basis for most contemporary MSIs) can depoliticize power relationships:

> [It] may neglect power inequalities among actors and regulatory frameworks, viewing the global public sphere

as a depoliticized arena of engagement among generic "stakeholders." The stark inequalities marking the local and international contexts that pose the most difficult governance challenges (e.g., those between transnational corporations, on the one hand, and workers and affected communities, on the other) translate into deep asymmetries among participants in multi-stakeholder arrangements and venues, from local consultations to global conferences.

Pointing to Fung and Wright's (2003) concept of "empowered participatory governance" (i.e., the basis for contemporary WSR initiatives), Rodríguez-Garavito calls for designing mechanisms that enhance the "countervailing power" of "rights holders, affected communities and civil society organizations" so that they can more effectively hold states and corporations accountable for implementing human rights norms and practices (2017a, 25). His call echoes that of the United Nations Special Rapporteur on Extreme Poverty and Human Rights, who has argued that people in poor communities not only have the right to participate in the design and implementation of programs that affect their lives but also have the right to hold duty-bearers to account, both public- and private-sector–based alike (United Nations General Assembly 2013, paragraphs 83–86).

These distinct conceptualizations of power are reinforced, in turn, through the research, training, and advocacy activities of organizations on each "side" of the MSI and WSR divide. Organizations in the MSI arena spearhead work informed by the Ruggie Principles (e.g., Baumann-Pauly and Nolan 2016)[13] while a growing academic literature critically analyzes whether and how MSIs can function to alter power relations in global supply chains (Anner 2018; Anner et al. 2013; Berliner, Greenleaf, Lake, Levi, and Noveck 2016) and critiques the broader normative foundations of the Ruggie Principles (Melish 2017; Deva and Bilchitz 2013). MSI-based organizations continue to develop training materials that reflect the reformist tendency, such as this excerpt from a 2014 assessment by SHIFT (the nongovernmental research group founded to work with business and governments on follow-up to the Ruggie Principles) of 43 corporate reports on stakeholder consultation processes. (The full report is analyzed in Chapter 3 of this book.) This excerpt is from the British-Australian extractive company Rio Tinto, which SHIFT highlights to illustrate advances in the development of evaluative criteria for stakeholder consultation:

> To maintain good relationships with communities, it is vital that the [production] site has formal processes for managing and, *where necessary*, escalating complaints to disputes and grievances . . . [C]omplaints, disputes, and grievances processes . . . should all include consultation with stakeholder groups to ensure that it meets their needs and that they will use it in practice. This includes facilitating community participation in resolution processes, *where appropriate* . . . To promote local awareness . . . the process is advertised in the local newspaper, site newsletters, community noticeboards and informally when [company] personnel visit the local community . . . the feedback procedure includes provisions for engagement and dialogue with the affected persons. . . . [The company] reports back to the community on how complaints are received and addressed [through a community forum]. (Rio Tinto 76, 79–80, as cited in SHIFT 2014a, Appendix B, point G, "Grievance Mechanisms"; emphasis added)

SHIFT offers this passage as an example of corporate best practice,[14] and its content reflects several key features common across MSI consultation mechanisms. First, the *company defines when a consultation is both "necessary" and "appropriate,"* and initiates the exchange with the community in response to complaints that have escalated to some unspecified level that requires the initiation of a dialogue. Second, there is an *implicit threshold (determined by the company) for how much suffering is acceptable* (or not) before dialogue is justified. The threshold is often based on a *hierarchy of rights* in which violations of physical integrity rights (such as the right not to be tortured, enslaved, or killed) rank as more deserving of attention than violations of economic rights—although stakeholders themselves often see the two problems as intrinsically linked and seek remedies for addressing both types of rights violations jointly.

The hierarchy of rights is an ongoing challenge in any negotiating setting. Legal scholars Knuckey and Jenkin (2015) explore the difficulty of balancing redress for varying types of rights violations in their analysis of a landmark case of stakeholder engagement in a Papua, New Guinea, gold mining site. At the Porgera Joint Venture Gold Mine, mine security personnel perpetrated widespread gender-based violence against women in the local community throughout the early 2000s (from roughly 2006 to 2012). Many of these women were poor informal miners themselves (people who scavenged around the edges of the mine for gold tailings) and, because they were engaged in an illegal activity as a means of economic survival, were doubly vulnerable to sexual abuse by mine security personnel (Human Rights Watch 2011; Columbia Law School et al. 2015).

Knuckey and Jenkin (2015) explore the Porgera case as a vehicle for understanding best practice in community grievance mechanisms. They contrast the company's consultation and remedy mechanisms developed for handling sexual violence complaints in the wake of the Porgera case with its conventional "operational grievance mechanisms" for handling ongoing "low-level" community complaints (Knuckey and Jenkin 2015, 801). The "low-level" violations include "employee complaints, property damage, and *relocation issues*," among other things (Knuckey and Jenkin 2015, 802; emphasis added). Only "grave or widespread human rights impacts and abuses" (Knuckey and Jenkin 2015, 802) could be handled through the new company-created remedy mechanism at Porgera.

But in failing to problematize the distinction between "low-level" and "grave" human rights abuses, Knuckey and Jenkin miss an opportunity to further refine the discussion of remedy—particularly when poverty is a driving force for women's economic survival strategies and their resulting criminalization in this case or others. At the most obvious levels, a, long-term remedy for sexual violation in the Porgera case would require support for employment alternatives for the women involved in informal, illegal gold mining activity. At a subtler level, the distinction between "high-level" and "low-level" human rights issues is complicated by the instrumental nature of some rights—such as the right not to be relocated without free, prior and informed consent, which is central to economic and cultural rights protection.

The potential for transcending the MSI and WSR divide hinges on recognizing the power differentials and the interdependence of human rights. Economic and social rights are more complex and multidimensional than the "high/low" distinction between rights implies. These rights often centrally undergird the concerns of workers and community members. (For example, community members relocated as a result of corporate activity would likely perceive relocation itself as more than a low-level concern—particularly indigenous peoples, whose right to cultural expression is intrinsically tied to control over land and natural resources.)

In Villa Altagracia and Bonao, community members not employed by Alta Gracia Apparel or Hanes consistently referenced their own needs for greater access to work and a broader base of economic security as integral to personal and community well-being. The deeply constrained economic survival choices made by women and men in the wake of industrial downsizing lead to community-wide vulnerability and, in Villa Altagracia, to extreme social violence reflected in the town's ranking as the most violent city in the Dominican Republic.

The alternative, argue proponents of the WSR model, is to shift power in global supply chains and surrounding communities well beyond MSIs. Partnerships between workers and community-based organizations function as catalysts of social and economic change in this model. In the United States, for example, worker centers have supported agricultural workers and other low-wage service workers (such as janitors or food service workers) in preventing wage theft and abusive working hours. In sectors such as construction and domestic work, the goal has been to build workers' capacity to take a more formal role in monitoring federal, state, and local labor laws—including by gathering evidence of violations, making claims, and designing proactive strategies for enforcement. Fine and Gordon's research on civil society–based co-enforcement of labor and economic rights

(2010, 559–562; Fine 2017) and writing by advocates and organizers (Albisa 2018, 2011; Asbed and Sellers 2013) detail the WSR model's "proven potential to afford protection for the most vulnerable and lowest-wage workers in global supply chains" (Worker-Driven Responsibility [WSR] Network n.d.). Outcomes include concrete changes in wage- and industry-specific regulatory standards, legal victories for victims of widespread labor rights abuse, and gains in community well-being.

Within the US agricultural sector, for example, the Fair Food Program for certifying tomato and other agricultural production in the American fast-food industry is the result of 17 years of struggle for protection against human trafficking and for guarantees of humane working conditions and living wages in the agricultural sector (Gordon 2017). Asbed and Sellers (2013) detail the emergence of the Coalition of Immokalee Workers—protagonists in the Fair Food Program—through which tomato pickers themselves exposed slave labor and widespread violations of working conditions in Florida's agricultural sector and have developed a "Fair Food Code of Conduct" and corresponding worker-driven monitoring, auditing, enforcement, and ongoing education practices that are now standard across this segment of the supply chain in this sector (see also Bauer 2016, 175–180). Similar struggles in the New England dairy sector have resulted in the Milk With Dignity certification program pioneered in Vermont (Gordon 2017; Damico and Sellers 2018). Within the light manufacturing sector, the Alta Gracia factory's governance model and the Accord on Fire and Building Safety in Bangladesh (Anner 2018; Gordon 2017; Donaghey and Reinecke 2018) are also characterized by the Worker-Driven Social Responsibility

Network as examples of worker-driven social responsibility models—and it is here that the interesting overlap emerges.

Each of these examples was cited by multiple participants in the 2017 University of Connecticut conference as evidence of successful consultation, though not always by way of dismissing the MSI model (Leipziger 2018, 8–11, 20–21). The juxtaposition of the MSI and WSR approaches has emerged organically in academic and policy dialogues on BHR—in some cases pitched, in others more subtly. This book has explored the potential and limits of both approaches to stakeholder dialogue. Across multiple industry sectors, companies continue to spend millions of dollars and hours on conventional supply chain monitoring and Ruggie-inspired MSI dialogues globally. At the same time, WSR-based approaches (which take intensive resources, time, and capacity-building to bring to fruition) have emerged as a vehicle for a "new social contract" rooted in "collective solutions built by and for communities"—both phrases integral to the title of report and platform for action released in May 2018 by coalitions central to the WSR movement (Albisa and Palmquist with Scott, Sellers, and Farr 2018).

If the MSI strand represents "what is" and the WSR strand represents "what could be," then communities like Villa Altagracia and Bonao demonstrate the challenges of implementing either approach in practice. How do people whose principal social currency in past models of supply chain management has been their story as "victims" (see Chapter 2) become agents in processes that transform the incentives of those people with power in supply chains? How can community members draw on the power of their own creative ideas to shape systems for safeguarding individual and community well-being in the context of globalizing business environments? Can a more dynamic notion of remedy—backward- and forward-looking—emerge from the findings in this book?

In this last section of this chapter, I draw on insights from systems engineering in order to explore approaches to advancing change in "failure-prone" systems such as the global light manufacturing sector, where economic rights violations persist. The conceptual foundations laid out will echo the dynamic "thinking-outside-the-box" perspective of the young artist I interviewed in Villa Altagracia,[15] who argued that a globalized future need not be a race to the bottom in terms of wages and working conditions but could be a journey to a new way of generating income that reflects new forms of knowledge, new ways of relating to other people, and new ways of projecting one's own creativity within a new form of economy.

MOVING FORWARD ON SEPARATE PATHS? GROUNDED THEORY ON COMPLEX PROBLEMS

Systems engineering scholar Krishna Pattipatti's work on problem-driven theory building (2017) and decision-making under uncertainty is grounded in an approach to solving seemingly intractable problems. The challenge of decision-making under uncertainty is integral to the task of developing new policies for promoting human rights that bridge "what is" (MSIs) with what "could be" (WSR approaches). First, while Pattipatti acknowledges that some problems seem both intractable and highly complex, breaking those problems down into smaller analytical challenges is the key toward solving them. Doing so requires (1) diagnostic data analysis (i.e., analysis of what happened and why) and (2) predictive analysis (i.e., analysis of what could happen).

The research design central to this book has aimed at both diagnosis and prediction. Drawing on historical data in Chapter 2 and large-N data in Chapter 3, the book assesses the emergence and evolution of stakeholder dialogue across countries and industries. The use of multiple forms of data at multiple levels of analysis helps break the complex issue of stakeholder dialogue into its constituent parts and map the scope of the process (What is it? Where is it happening? Why is it working or not?). Qualitative interview data gathered at the local level, in turn, reveal the seemingly intractable set of factors that stymie successful community engagement in light manufacturing towns like those where the fieldwork central to this book took place (Chapters 4 and 5).

Second, Pattipatti identifies the difficulty of valuing systems with a high propensity for failure. In these settings, "exact" solutions are not practical. Instead, the task is to identify the causes of problems and adjust solutions to address aspects of the problem incrementally (i.e., evaluating through approximation). The key is to focus on how to constantly upgrade information as the problem evolves in order to adjust our responses. Accurate metrics of the depth and severity of the problem are essential, but finding them requires decomposition of the problem itself into its constituent parts. As the evidence in Chapter 3 and this chapter makes clear, contemporary supply chain management systems have a high propensity for failure and stakeholder dialogue practices often fail to engage a broad range of community members substantively.

Within the domain of stakeholder dialogue policies, then, WSR approaches offer a vehicle for engaging workers themselves in identifying the nature and scope of problems and in teasing out the interconnections between interdependent rights. In this book, micro-level data gathered at the community level through interviews have helped identify not only the obstacles that prevent successful stakeholder dialogue in both the Villa Altagracia and Bonao, but also the potential for dynamic and constantly adaptive solutions. The key to implementing dynamic and adaptive solutions is long-term commitment to the task. Generating this type of commitment is easier in the laboratory and industrial settings in which systems engineering scientists work than in the messy reality of towns where light industry is prevalent, sunk costs are low, and the threat of corporate exit looms large.

Literature on labor in the developing world has grappled centrally with the disconnect between practical reality and idealized policy design, and places special emphasis on the right that vulnerable people have to participate in the design of remedy, in particular. Susan Bissell (2005, 385), a scholar with extensive policy experience and field research on child labor in Bangladesh's garment industry, observes that the "existing strategies of vulnerable populations need to be factored into understandings, which in turn inform policy . . . [in order to give] substance to claims of 'participation,' not only in the research but in the development of future courses of policy and action." Bissell cautions that "assumed passivity and lack of agency" on the part of people "deny both the reality of their daily lives and their human rights" (2005, 385). The challenge, then, in formulating remedy for human rights abuse experienced by people in settings such as the global textile industry is "to understand, face, and deal with *what is* while at the same time addressing *what ought to be*" (2005, 385; italics in the original).

Bissell's work and my own have both been informed by study of child labor in Bangladesh's garment industry (Hertel 2006, 31–54), a problem that long predated the collapse of the Rana Plaza factory in 2013. Child labor in Bangladesh led to the first wave of consumer-driven outrage over labor conditions in that country in the early 1990s. Following a protracted transnational campaign and resulting pushback at the national level in Bangladesh, a Memorandum of Understanding (MOU) was ultimately brokered in 1995 among representatives of the Government of Bangladesh; the major industry association for garment export; the International Labour Organization (ILO); and the United Nations Children's Fund (UNICEF). The MOU included provisions for schooling and stipends for children identified as having worked in factories during a specific period of time. The MOU was thus a precursor to the 2013 Accord on Fire and Factory Safety in Bangladesh. Through the hard-fought participation of South Asia–based human rights advocates (such as Nobel laureate Kailash Satyarthi), a new ILO Convention 182 on the Worst Forms of Child Labor emerged in 1998 during this earlier wave of supply chain advocacy in Bangladesh.

Critically, the child labor MOU did not involve workers centrally in monitoring child labor; the Accord on Fire and Factory Safety in Bangladesh *does* involve worker monitoring of safety conditions. The child labor MOU nevertheless sought a broader scope of remedy for children removed from factories (i.e., schooling and stipends) and in the process committed government and private-sector actors together to address some of the "push" factors underlying child labor in the community (i.e., poverty, lack of schooling alternatives, or other social and economic factors). Ultimately, only a quarter of the original children identified through enhanced factory inspections (i.e., roughly 10,000 out of an estimated 40,000 to 50,000) could be located for support (Hertel 2006, 35, 47). Two decades later, the Accord on Fire and Factory Safety in Bangladesh legally obliges brands to shoulder the cost of factory repairs rather than pushing them down the line to workers in the form of lowered wages, thus helping to address structural pressure on low wages. Connecting these two waves of problems and policies in Bangladesh serves to spotlight the evolution of more structurally grounded remedies in light manufacturing. It helps bridge the MSI and WSR divides because the evolution of remedy in this country setting demonstrates how "incremental progress" and "adjusting solutions" takes place, which as Pattipatti argues is central to change in failure-prone contexts.

By focusing on the skills, abilities, needs, and rights of people in surrounding communities as a source of knowledge and power, this book adjusts our sources of information in ways that alter the concept of remedy. This adaptive orientation to stakeholder dialogue reveals the roots of structural inequality in an effort to preempt continued and deepening economic and social divides along the global supply chain. As Ackerly argues, exposing the structures is not enough; taking steps toward just remedy is necessary (albeit incremental) if BHR policies such as stakeholder dialogue are to have an impact on people's lives at the local level. Building the capacity of communities to engage companies and their own governments in moving beyond the impasse of "what is" is the first step toward more effective and inclusive dialogue.

Epilogue

EVERY DAY, EACH OF US eats, wears, and uses products made by people we will never see or meet. Our fates are tethered—not only environmentally (think of the risks to consumers of unsafe products and the health hazards to the people who make them) but also economically. Downward pressure on wages and working conditions in one location has repercussions in homes and communities well beyond it. The scope of global manufacturing and the speed with which companies can shift production tie our fates even more closely.

The experiences of people who live in manufacturing towns in the Dominican Republic are not delinked from those of the people in places where corporations are headquartered abroad. They are connected. Philanthropic efforts (like Hanes's support for medical services and capacity-building, discussed in the Bonao case study) reveal one level of connections, but worker-driven social responsibility (WSR) efforts (like those involved in creating and sustaining production at Alta Gracia Apparel) entail transnational alliances on an even higher order of magnitude. Scores of visits by Dominican workers to American colleges and universities have helped to create a market for "living wage" garments, and intensive ongoing transnational support helps sustain local monitoring, collective bargaining, and production improvements. But the Alta

Gracia factory is a small one in a much bigger sea of conventional manufacturing.

At the local level in communities like these and beyond, the economically vulnerable condition of people *beyond* the factory floor is less visible. We don't often hear their complex and varied views on who is responsible for community well-being, nor on companies' or governments' role in safeguarding rights. This book aimed to bring such voices to bear on academic and policy debates over the nature of "remedy" in the evolving field of business and human rights (BHR). Moving beyond the ritual of consultation toward substantive community participation in the design and implementation of BHR policies is a first step. Changing the economics of supply chains so that retailers and brands bear their fair share of the cost of a living wage is a next step. Changing the incentive structures in supply chains themselves to create a demand for ethically produced goods is a third step (Hertel, Scruggs, and Heidkamp 2009; Scruggs, Hertel, Best, and Jeffords 2011).

But contemporary supply chain governance remains prone to failure, not only on the part of companies but also on the part of governments. If it remains largely focused on damage control and punitive in nature, with the costs of compliance principally passed on to workers in the form of lower wages, then the fates of people tied up in global supply chains (all of us) will become increasingly mutually insecure. In manufacturing communities pitted against one another globally, a fuller interpretation of "remedy" necessarily means engaging people at the grassroots level as allies in designing solutions for workplace-based and community-based problems that cripple not only productivity but also equity.

Workers and community members at both ends of the supply chain want to be heard, individually and collectively. Recognizing their agency means moving beyond the assumption that they are passive or uninformed about what they need to "live the kind of life we have reason to value," to quote Nobel Prize–winning economist Amartya Sen (cited in Soundararajan, Brown, and Wicks 2016, 31). Many possess the types of fine-grained detail critical to adapting decision-making in failure-prone settings and to adjusting solutions. Getting beyond "what is" toward "what could be" the scope and content of stakeholder dialogue across industries and geographies thus requires thinking beyond present practices. WSR systems offer a first valuable step in this direction. Through multiple forms of research, this book has offered insights into how to foreground related principles of participation, transparency, and accountability.

The genealogy of community consultation laid out in Chapter 2 is a starting point for chronicling the evolution of stakeholder dialogue. New sources of historical data—from oral histories to archives—will continue to emerge, and the interpretation here marks but a first take at mapping change from the 1970s to the present in consultation practices. Other scholars can and will extend the timeline backward and forward and will layer on new forms of evidence, bringing new voices to the fore as we build a richer history of the struggle to fulfill economic rights over time.

In a more quantitative vein, the book has laid out provisional findings on stakeholder spread and depth based on newly available methods for using the Business and Human Rights Resource Centre data featured in Chapter 3. Ideally, other researchers will build upon and advance scholarly and policy discussions on stakeholder dialogue using this powerful resource. The book also draws on qualitative interview data based on comparative case studies (in Chapters 4 and 5) in an effort to take stock of what works and what does not from the perspective of community members themselves. For many of the people I spoke with in the Dominican Republic, being interviewed was a new experience, but their stories transcend time and space. There are millions of communities globally where similar qualitative research could take place worldwide. The key to its being richly informed is the generosity of people at the grassroots level in sharing their stories, along with the resourcefulness of nongovernmental organization and government representatives who are willing to gather and share other forms of data. The policy pathways laid out in Chapter 6 provide a series of signposts on the way forward, not a roadmap. Hopefully, people from multiple professional and regional spaces will make contributions to untangling the tethered fates of companies, communities, and rights at stake using the findings presented here and the subsequent research they help inform.

APPENDIX 1

INTERVIEW QUESTIONNAIRE

Principal Investigator: Dr. Shareen Hertel (Associate Professor, Univ. of Connecticut)

Contact Information: Department of Political Science
341 Mansfield Road Unit 1024
Storrs, CT 06269-1024 USA
Email: shareen.hertel@uconn.edu

Study Title: Community Engagement in the Textile Industry: Dominican Republic Field Study

1. Can you tell me about your own work history?
2. How would you describe the employment opportunities in the community where you live?
   a. Have these opportunities changed over time? Why or why not?
3. How would you describe the level of social well-being for people in the community where you live?
   a. Has this changed over time? Why or why not?

4. Have you or anyone you know worked in the textile industry?
   a. If so, where?
   b. If so, for how long?
   c. If not, why not?
5. How do you get information about company policies or practices in your community? (Choose those answers that apply from the list below.)

   a. ____ From what I see or hear personally. Give an example: _____
   b. ____ Through the formal media, such as newspapers, TV, radio. (Circle those that apply.)
   c. ____ Through social media. Give an example: __ _____
   d. ____ From materials the company distributes publicly. Give an example: _____
   e. ____ Other ways? Give examples: _____

OR

   f. ____ I don't have any information about company policies or practices in this community. (If not, why not?)
6. In your opinion, who is principally responsible for the well-being of people in the community where you live, and why? (Choose those answers that apply from the list below.)
   a. ____ People themselves who live in this community. Why?
   b. ____ The companies in this community. Why?
   c. ____ Our government. Why?
   d. ____ NGOs or churches. Why?
   e. ____ Someone (or something) else. Why?

7. Occasionally, people from the community meet with company representatives. Has that ever happened in the community where you live? Tell me about it.
8. If you were invited to participate in a dialogue with representatives of a company, would you participate?
   a. YES or NO
   b. Why?
9. If you did participate, then:
   a. What would your top priority be?
   b. What kind of information would you need to make the conversation effective?
   c. What kinds of resources or skills would you need to make the conversation effective?
10. Before we close, can you tell me if you agree or disagree with the following statements:
    a. Companies do not have a social obligation to people beyond their employees or shareholders.
    b. If a company's leadership chooses to make a contribution to the community (such as support for a local school or clinic), community members should accept it and not expect more.
    c. Labor unions can only help their members, not people outside the union.
11. Are there other people in the community where you live who would want to discuss the kinds of issues we've talked about today?
    a. Why, or why not?
12. Is there anything else you'd like to discuss with me?

APPENDIX 2

CUESTIONARIO

Investigador principal: Dr. Shareen Hertel (Prof. asociada, Universidad de Connecticut)
Información de contacto: Departmento de Ciencias Políticas
341 Mansfield Road Unit 1024
Storrs, CT 06269-1024 USA
Email: shareen.hertel@uconn.edu
Título del estudio: Participación de la comunidad en la industria textil: estudio de campo en la República Dominicana

1. ¿Podría hablarme de su historial de trabajo?
2. ¿Cómo describiría las oportunidades de trabajo en la comunidad en donde usted vive?
   a. Estas oportunidades ¿han cambiado con el tiempo? ¿Por qué o por qué no?

3. ¿Cómo describiría el nivel de bienestar social de las personas que residen en la comunidad en donde usted vive?
   a. ¿Esto ha cambiado a través del tiempo? ¿Por qué o por qué no?
4. ¿Usted, o cualquier persona que usted conozca, ha trabajado en la industria textil?
   a. De ser así, ¿en dónde?
   b. De ser así, ¿por cuánto tiempo?
   c. De no ser así, ¿por qué no?
5. ¿Cómo obtiene información sobre las políticas o prácticas de la compañía en su comunidad? (Escoja aquellas respuestas que apliquen de la lista a continuación)

   a. ____ De lo que veo o escucho personalmente. Dé un ejemplo: _____
   b. ____ Mediante medios de comunicación formales, como periódicos, televisión, radio (circule aquellas que apliquen)
   c. ____ Mediante las redes sociales. Dé un ejemplo: _____
   d. ____ De materiales que la compañía distribuye públicamente. Dé un ejemplo: _____
   e. ____ ¿De alguna otra forma? Dé ejemplos: ____ _____

O

   f. ____ No tengo ninguna información sobre políticas o prácticas de la compañía en esta comunidad. (¿Por qué no?)
6. En su opinión, ¿quién es la persona responsable por el bienestar de las personas en la que usted

vive? ¿Por qué? (Escoja aquellas respuestas que correspondan de la siguiente lista)
   a. ____ La misma gente que reside en esta comunidad. ¿Por qué?
   b. ____ Las compañías en esta comunidad. ¿Por qué?
   c. ____ Nuestro gobierno. ¿Por qué?
   d. ____ Las ONGs o iglesias. ¿Por qué?
   e. ____ ¿Alquien (o algo) más? ¿Por qué?
7. Ocasionalmente, gente de la comunidad se reúne con los representantes de las compañías. ¿Alguna vez ha sucedido esto en la comunidad en donde usted vive? Hábleme al respecto.
8. Si a usted se le invitara a participar en un diálogo con un representante de alguna compañía, ¿participaría?
   a. SÍ o NO
   b. ¿Por qué?
9. Si usted participa, entonces:
   a. ¿Cuál sería su prioridad número uno?
   b. ¿Qué tipo de información usted necesitaría para hacer la conversación una efectiva?
   c. ¿Qué tipos de recursos o habilidades usted necesitaría para hacer de la conversación una efectiva?
10. Antes de terminar, ¿podría decirme si usted está de acuerdo o en desacuerdo con las siguientes oraciones?:
    a. Las compañías no tienen una obligación social hacia las personas más allá de sus empleados o accionistas.
    b. Si un líder de la compañía decide hacer una contribución a la comunidad (tal como

proporcionar ayudas para una escuela o clínica local) los miembros de la comunidad deberían aceptarlo y no esperar nada más.
  c. Los sindicatos solo pueden ayudar a sus miembros, no a aquellas personas fuera de ellos.
11. ¿Hay otras personas en la comunidad en donde usted vive a quienes le gustaría discutir los tipos de temas de los que hemos hablado hoy aquí?
  a. ¿Por qué, o por qué no?
12. ¿Hay algo más que usted quiere discutir conmigo?

NOTES

Chapter 1

1. See articulated in the United Nations Guiding Principles on Business and Human Rights (United Nations Human Rights Council 2011, 27–35) or "Ruggie Principles," named for their architect John Ruggie, the United Nations Secretary General's representative on BHR from 2005 to 2011.
2. Backer (2012) has argued that duty to remedy as articulated in the Ruggie Principles lacks an independent "normative justification" (212, 150) from either the state duty to protect rights or the corporate duty to respect rights (i.e., the other two pillars of the Ruggie Principles). He warns of the potential for interpretative "fragmentation" of the Principles more generally (2012, 157).
3. Notably, the United Nations Guiding Principles on Business and Human Rights "ask companies to adhere to an international standard in their operations—thus moving beyond local law—hence a theoretical race to the top" (Ramasastry 2015, 246).
4. One exception is Fulton, Ha, Karimian, Lerner, Meier, and Plessis (2015).
5. See, for example, a "Stakeholder Dialogues Manual" produced by Germany's semiautonomous development agency Deutsche

Gesellschaft fur Internationale Zusammenarbeit/GIZ (2011) and a best practices survey on "Operational Grievance Mechanisms" produced by the International Petroleum Industry Environmental Conservation Association/IPIECA (2012).

6. In the US legal system, moreover, poor people do not have the status of a "protected class," which means they are unable to seek collective damages through class-action litigation unless they can demonstrate the willful denial of access to economic rights based on ascriptive characteristics such as race or gender (Albisa 2011).

Chapter 2

1. Cutler similarly takes an historical approach to explain how corporate hegemony has evolved through rule-making on international commercial relations (1999). Her analysis focuses on elite actors, institutions, and processes, whereas this chapter focuses on popular-level dynamics and implications.
2. On colonial patterns of extraction, see Rodney (1972).
3. Kingstone (2011) provides region-specific analysis of these broader trends.
4. For the IMF's discussion of the launch of PRSPs, see https://www.imf.org/external/NP/prspgen/review/2002/032602a.pdf (accessed April 8, 2017).
5. Roughly 95% of all "national regulations related to foreign direct investment were modified" during this decade (i.e., from 1991 to 2001; see Ruggie 2013, xxv).
6. For information on Romano, see the Romano Archives at the University of Connecticut (http://archives.lib.uconn.edu/islandora/object/20002%3A20110094, accessed December 19, 2017) and Zack (2015). See discussion of related child rights advocacy in Hertel (2006). For Terry's work, see The Aftermath Project (http://theaftermathproject.org/story), accessed December 19, 2017.
7. See Kaeb and Scheffer (2011; 2013). See also Collingsworth (2003).

8. Details on the full survey are available through the Boston Consulting Group (http://sloanreview.mit.edu/projects/joining-forces/?utm_source=BCG&utm_medium=referral&utm_campaign=susrpt14, accessed March 8, 2017).
9. Labor union density throughout the industrialized world is at all-time low levels, making this type of consultation increasingly irrelevant for a vast majority of workers in skilled, unskilled, and informal-sector occupations alike.

Chapter 3

1. Barkemeyer (2009) analyzed 400 case studies derived from reporting by member companies of the United Nations Global Compact and found that the "vast majority of efforts" at promoting the corporate social responsibility "are targeted at Western home markets and focus on environmental issues, rather than corruption, labor standards, or human rights" (Barkemeyer 2009, cited in Mwangi, Reith, and Schmitz 2013, 214).
2. The samples central to several of the studies discussed are limited to under 50 companies and the authors of those studies also caution against overly generalizing from their findings (see SHIFT 2014a and 2014b; Collins, Evans, Hung, and Katzenstein 2017). Another study analyzed in this chapter draws on 250 grievance reports filed under BHR reporting procedures of the Organisation for Economic Cooperation and Development (OECD), but OECD membership is not universal, so the findings extrapolated from such reporting are necessarily qualified as well; see Daniel, Wilde-Ramsing, Genovese, and Sandjojo (2015).
3. See http://bhrrc.org/ (accessed May 25, 2018).
4. Ariana Javidi and Emily Kaufman carried out this initial hand-coding in 2015 and 2016, and I acknowledge their contribution here. Notably, the BHRRC revised its online search categories subsequent to the initial hand-coding, rendering it impossible to replicate those results. I summarize the hand-coding procedure in this footnote partly

for historical value but note that it aligns with the results of computer coding described in the body of this chapter. Moreover, it offers a template for hand-coding even with the current BHRRC revised search categories, in light of the ongoing developmental state of the API. In wave one of the hand-coding, we varied the mix of categories over three successive rounds of searching: within round one, in order to winnow our universe of cases to those most relevant to this project, we coded for "Industry Name" (extractive, pharmaceutical, water, and light manufacturing) together with regional delimiters (Asia OR Latin America) and mode of consultation ("stakeholder," "community," "indigenous," and/or "consultation"). The second round of coding within wave one was for "Issues," and within that overarching category, we analyzed data from the "Labor" category using the following search terms: "labor general," "living wage," and "child labor" plus the above four terms (i.e., stakeholder, community, indigenous, consultation). The final round within wave one of coding integrated terms from the first two rounds with new terms in the "Company Policy/Steps" category, including "company advances," "steps," "reporting," and "other." During wave one, we thus detected only references to reporting on region-wide trends by industry sector in relation to consultation on labor issues. In wave two, we narrowed the search by a specific subset of countries within each region in order to ensure that we were not missing sub-regional anomalies; these results confirmed the findings in wave one. We then conducted a third wave of coding (repeating all steps above) in order to ensure intercoder reliability; wave three revealed no substantive discrepancies between coders.
5. Rajeshwari Majumdar has contributed centrally to the pilot phase of testing BHRRC's new API and the analysis in this chapter is enriched by her work. Coding parameters are available upon request from the author.
6. The first and latest version of the API as of May 2018 is v1.
7. We performed our data collection between May 10, 2018, and May 15, 2018.

8. Among "Stories" that are collections of items, some had a comprehensive set of "Categories" linked to the main "Story" that we could extract as we would from a standalone "Story." However, some did not have any "Categories" linked to the main "Story" but rather had "Categories" linked to one or more of the "Components" within the "Story." To incorporate them into our analysis, we extracted the set of "Categories" linked to each "Component" within the "Story" and took the union of all the resulting sets to produce a comprehensive set of "Categories" for the "Story." We dropped all "Stories" that did not have any "Categories" whatsoever as well as all "Stories" that had more than 50 "Categories" to decrease the risk of type 1 errors, or false positives ("Stories" that have that many "Categories" are typically entire journal issues or annual United Nations reports that incorporate articles and documents related to many disjoint events in the collection). Of the 11,785 "Stories" on the BHRRC's website, 635 either did not have any "Categories" or had too many "Categories," so we report findings in this chapter on the remaining universe of 11,150 units of data.
9. The BHRRC's list of "Categories" contains 192 countries; we removed the Cook Islands, the Vatican, and Yugoslavia from consideration. A full list of countries is available at https://www.business-humanrights.org/en/search-topics (accessed May 25, 2018).
10. The total number of possible combinations is the product of 189 (countries), 11 (industries), and 3 (mechanisms).
11. Although the BHRRC includes a "Region" search category, the sub-list of countries in each region is not intuitively linked to continents, nor are Latin American countries grouped to include Mexico, Central and South America, and the Caribbean. So, we created regional groupings of Asia, Africa, Europe, Oceania, Latin America, and USA/Canada, and we associated individual countries with each region accordingly.
12. OECD Watch is a network of 100 civil society organizations based in over 50 countries that together monitor corporate misconduct. The 2015 report analyzes the functioning

of the National Contact Points system established by the OECD to promote enforcement of the OECD Guidelines for Multinational Enterprises (adopted in 1976 and revised most recently in 2011 following the release of the Ruggie Principles), see Daniel, Wilde-Ramsing, Genovese, and Sandjojo (2015, 52), hereinafter OECD Watch (2015).

13. SHIFT notes that the sample of companies analyzed was not intended to be representative but was composed of companies with a "known record of having formal human rights programs and/or strong disclosure of environmental, social, and governmental performance" (SHIFT 2014a, 5). This report was the second of two major SHIFT studies on the corporate responsibility for remedy and related grievance mechanisms; see also SHIFT (2014b).
14. To access the full methodology document, see https://msi-database.org/data/Project%20Methodology%20and%20Classification%20Guide%20-%20MSI%20Database.pdf (accessed September 29, 2017).
15. This information and other census data are accessible through the República Dominicana Sistema Interactiva de Consulta Censo (SICEN; 2010): http://sicen.one.gob.do/ (accessed December 12, 2017).
16. The field study was covered under University of Connecticut Institutional Review Board Protocol #H16-294, which required signed consent on the part of people interviewed and in turn guaranteed their anonymity in the context of any future published work.
17. This segment of data gathering was also covered under an amendment to the same Institutional Review Board Protocol #H16-294.
18. For a similar methodological approach to qualitative data collection and analysis, applied in another research setting, see Godoy 2018.

Chapter 4

1. Interview VAG022060117 on June 1, 2017.
2. Interview VAG0115060117 on June 1, 2017.

3. Interview VAG008053117 on May 31, 2017.
4. These wage figures are based on a 2010 living wage calculation developed by the WRC based on its original market basket study, which also accounts for the Dominican Republic government's mandatory deductions from salaries for both health care and retirement (2010, 6). The WRC used an Interbank exchange rage as of June 24, 2010; US Department of the Treasury (Bureau of the Fiscal Service) historical data show that this conversion rate was 36.7 Dominican Republic pesos (RDP) to 1 US dollar (USD) as of June 30, 2010; see https://www.fiscal.treasury.gov/fsreports/rpt/treasRptRateExch/0610.pdf (accessed January 8, 2018).
5. Adler-Milstein and Klein distinguish between stakeholders and workers, with the former including manufacturers, brand representatives, and university administrators (2017, 38, 186).
6. When I report prices quoted to me by respondents I interviewed in summer 2017, I use the US Department of the Treasury (Bureau of the Fiscal Service) conversion rate of 47.6 Dominican Republic pesos (RD pesos) to 1 US dollar (USD) reported by the US Department of the Treasury (Bureau of the Fiscal Service) on September 30, 2017 (https://www.fiscal.treasury.gov/fsreports/rpt/treasRptRateExch/currentRates.htm; accessed January 8, 2018). For an updated calculation of living wage by the Fair Labor Association (2016), see http://www.fairlabor.org/sites/default/files/dominican_republic_benchmarks_and_charts_072816.pdf)accessed January 8, 2018). For more general macroeconomic data, see the International Monetary Fund World Economic Outlook indicators for the Dominican Republic (projections downloaded as of January 8, 2018): https://www.imf.org/external/pubs/ft/weo/2016/01/weodata/weorept.aspx?sy=2014&ey=2021&scsm=1&ssd=1&sort=country&ds=.&br=1&pr1.x=33&pr1.y=13&c=243&s=NGDPPC%2CPPPPC%2CPCPI%2CPCPIPCH%2CPCPIE%2CPCPIEPCH%2CLUR%2CLP&grp=0&a=.
7. The field study was covered under University of Connecticut Institutional Review Board Protocol #H16-294, which

required signed consent on the part of people interviewed and in turn guaranteed their anonymity in the context of any future published work.
8. Schrank's assessment is based on analysis of a subset of workers surveyed in 2004 in the Dominican Republic during the fourth in a series of nationally based public opinion polls conducted by researchers at the Pontifícia Universidad Católica Madre y Maestra and known as "La Encuesta Nacional de Cultural Política y Democracia" or DEMOS series (i.e., National Survey of Political Culture and Democracy)—see Brea, Duarte, and Seligson (2005). Full results and analysis of DEMOS polling data from multiple years are available through the Latin American Public Opinion Project (LAPOP) of Vanderbilt University (http://www.vanderbilt.edu/lapop/dominican-republic.php; accessed February 7, 2017).
9. Interview VAG005053117 on May 31, 2017.
10. Interview VAG004053117 on May 31, 2017.
11. Interview VAG002053117 on May 31, 2017.
12. Interview VAG022060117 on June 1, 2017.
13. Interview VAG010053117 on May 31, 2017.
14. Interview VAG017060117 on June 1, 2017.
15. Interview VAG004053117 on May 31, 2017.
16. Interview VAG012060117 on June 1, 2017.
17. Interview VAG013060117 on June 1, 2017.
18. Interview VAG017060117 on June 1, 2017.
19. Interview VAG025060117 on June 1, 2017. Notably, well over half (i.e., 56%) of the population regularly participates in religious organizations, and Dominicans report relatively high and consistent levels of religious participation regardless of demographics (Brea et al. 2006, 151–152, 157), along with deep political party affiliations.
20. Interview VAG027060117 on June 1, 2017.
21. Interview VAG018060117 on June 1, 2017.
22. Interview VAG014060117 on June 1, 2017.
23. Ibid.
24. Interview VAG020060117 on June 1, 2017.
25. Interview VAG019060117 on June 1, 2017.
26. Interview VAG010060117 on June 1, 2017.
27. Interview VAG004053117 on May 31, 2017.

28. Interview VAG021060117 on June 1, 2017.
29. Interview VAG009053117 on May 31, 2017; same phrase also invoked in interview VAG027060117 on June 1, 2017.
30. Interview VAG013060117 on June 1, 2017.
31. Interview VAG007053117 on May 31, 2017.
32. Interview VAG004053117 on May 31, 2017.
33. Interview VAG022060117 on June 1, 2017.
34. Interview VAG024060117 on June 1, 2017.
35. Interview VAG003053117 on May 31, 2017.
36. Interview VAG002053117 on May 31, 2017.
37. Interview VAG08060117 on June 1, 2017.
38. Interview VAG014060117 on June 1, 2017.
39. Interview VAG015060117 on June 1, 2017.
40. Interview VAG016060117 on June 1, 2017.
41. Interview VAG021060117 on June 1, 2017.
42. Interview VAG027060117 on June 1, 2017.
43. Interview VAG011053117 on May 31, 2017.
44. Her comments on the citrus refinery were also salient from an environmental rights perspective: "For years, they've been planting oranges, nothing more . . . without giving the earth an opportunity to revive. You know all the chemicals that they put into these trees, these bushes and plants? They have damaged the environment. Moreover, the code of conduct they've made? They ought to draft another with provisions for care of the land because they're just renting the land—it's not theirs; these lands belong to be Villa Altagracia" (interview VAG027060117 on June 1, 2017).
45. Interview VAG003053117 on May 31, 2017.
46. Ibid.
47. Interview VAG009053117 on May 31, 2017.
48. Interview VAG004053117 on May 31, 2017.
49. Interview VAG024060117 on June 1, 2017.
50. Interview VAG027060117 on June 1, 2017.
51. Interview VAG012060117 on June 1, 2017.
52. Interview VAG023060117 on June 1, 2017.
53. Interview VAG026060117 on June 1, 2017.
54. Interview VAG027060117 on June 1, 2017.
55. Interview VAG008053117 on May 31, 2017.

Chapter 5

1. Interview BON002060517 on June 5, 2017.
2. Interview BON006060617 on June 6, 2017.
3. This and other census data are accessible through the República Dominicana Sistema Interactiva de Consulta Censo (SICEN 2010): http://sicen.one.gob.do/ (accessed March 1, 2018).
4. This field study was covered under University of Connecticut Institutional Review Board Protocol #H16-294.
5. Interview BON001060517 on June 5, 2017.
6. This relative weighting of the textile sector's significance over extractives was borne out locally in interviews. As a retirement-aged woman observed: "The majority of these young people work in *la zona* [the export processing zone]. Many other young men work in Falconbridge, in Barrick Gold—there is a lot of work there. But [there's] more work in the free-trade zone" (interview BON009060717 on June 7, 2017).
7. Interview BON016060717 on June 7, 2017.
8. Interview BON011060717 on June 7, 2017.
9. Ibid. The community strike referenced by this woman in Las Delicias was similar to one referenced in an interview I conducted in Villa Altagracia, where a young man explained that a community strike there had resulted in the construction of a bridge; see interview VAG002053117 on May 31, 2017.
10. Interview BON006060617 on June 6, 2017.
11. This institution is not an orphanage in the classic sense of the word (only a few of the upwards of 40 girls, ages three to 18 years old, who lived in the multi-building complex were actually without parents) but rather a safe place for children whose families could not care for them, owing to financial, social, or emotional constraints.
12. As noted in Chapter 4, the adolescent birthrate nationally in the Dominican Republic is 99.6 per every 1,000 women aged 15 through 19 (United Nations Development Programme 2015, 6); with 22% of all births in the country occurring among women aged 15 through 19, the country's adolescent

birthrate is 34% higher than the regional average (United Nations Development Programme 2017, 12).
13. Interview BON003060517 on June 5, 2017.
14. Interview BON009060717 on June 7, 2017.
15. Interview BON007060617 on June 6, 2017, and interview BON015060717 on June 7, 2017. For media coverage of these efforts, see https://www.youtube.com/watch?v=AJ8FYHzWusE (accessed March 16, 2017).
16. Interview BON004060517 on June 5, 2017.
17. Data shared with the author by two senior INFOTEP administrators at the agency's headquarters in Santo Domingo on June 9, 2017 (memo on file with the author).
18. Interview BON008060617 on June 7, 2017; also referenced in the context of participant observation.
19. Interview BON016060717 on June 7, 2017.
20. As Schrank argues, "skilled labor inspectors and their associates" in the Dominican Republic have "not only blocked the proverbial low road by deterring repression and exploitation but *paved a potentially higher road* by building bridges between their private sector interlocutors and public educational, training, and financial institutions" (Schrank 2013, 300; emphasis added). Mosley and Uno (2007) have cautioned that a straight race to the bottom is an oversimplification of reality and that labor conditions in developing countries vary depending on the structure of export-oriented industry in a given country; better conditions are correlated, in their research, with higher-skill and higher-value-added export industries.
21. Interview BON005060617 on June 6, 2017.
22. Ibid.
23. Interview BON006060617 on June 6, 2017.
24. Interview BON012060717 on June 7, 2017.
25. Interview BON007060617 on June 6, 2017.
26. Interview BON006060617 on June 6, 2017.
27. Interview BON011060717 on June 7, 2017.
28. Interview BON015060717 on June 7, 2017.
29. Interview BON001060517 on June 5, 2017.
30. Interview BON008060617 on June 6, 2017.
31. Ibid.

32. Interview BON005060617 on June 6, 2017.
33. Interview BON011060717 on June 7, 2017.
34. Interview BON013060717 on June 7, 2017.
35. Interview BON005060617 on June 6, 2017.
36. Interview BON009060717 on June 7, 2017.
37. Interview BON002060517 on June 5, 2017.
38. Interview BON004060517 on June 5, 2017.
39. Interview BON006060617 on June 6, 2017.
40. Villa Altagracia has a similarly low density of local NGOs registered with MEPyD: only 20 registered organizations are located in that town out of the 293 registered in the state of San Cristóbal, where Villa Altagracia is located. Data on NGO density in both towns is based on lists shared with the author by a representative of MEPyD on May 26, 2017.
41. Interview BON006060617 on June 6, 2017.
42. Interview BON001060517 on June 5, 2017, and interview BON010060717 on June 7, 2017.
43. Interview BON009060717 on June 7, 2017.
44. Interview BON008060617 on June 6, 2017.
45. Interview VAG002053117 on May 31, 2017, and interview VAG026060117 on June 1, 2017.
46. Interviews VAG007053117, VAG008053117, and VAG006053117, all on May 31, 2017.
47. Interview VAG018060117 on June 1, 2017.
48. Interview BON008060617 on June 6, 2017.
49. Interview SD001052317 on May 23, 2017; see also Rodríguez (2017).
50. Interview BON012060717 on June 7, 2017.
51. Interview BON001060517 on June 5, 2017.
52. Interview BON003060517 on June 5, 2017.
53. Interview BON001060517 on June 5, 2017.
54. Interview BON004060517 on June 5, 2017.
55. Interview BON012060717 on June 7, 2017, and interview BON015060717 on June 7, 2017.
56. Interview BON013060717 on June 7, 2017.
57. Interview BON015060717 on June 7, 2017.
58. Interview BON013060717 on June 7, 2017.
59. Interview BON002060517 on June 5, 2017.
60. Interview BON011060717 on June 7, 2017.

61. Interview BON008060617 on June 6, 2017.
62. Interview BON011060717 on June 7, 2017.
63. Interview BON013060717 on June 7, 2017.
64. Interview BON016060717 on June 7, 2017.
65. Ibid.
66. Interview BON008060617 on June 6, 2017.
67. Interview BON006060617 on June 6, 2017; emphasis added.
68. Interview BON012060717 on June 7, 2017.
69. Interview BON010060717 on June 7, 2017.
70. Interview BON004060517 on June 5, 2017.
71. Interview BON005060617 on June 6, 2017.
72. Interview BON016060717 on June 7, 2017.
73. Interview BON005060617 on June 6, 2017.
74. Interview BON011060717 on June 7, 2017.

Chapter 6

1. Mark Anner's research demonstrates the confluence of these pressures on wage rates in Bangladesh (2018, 9).
2. Some of the earlier literature on stakeholder consultation anticipated related challenges (Fransen and Kolk 2007, 671, 673, 676) and drew on the public policy literature to suggest concrete strategies for dispute resolution, including "encouraging joint fact-finding; committing to work with affected stakeholders to minimize or correct adverse project-related impacts; accepting responsibility, admitting mistakes, and sharing power; acting in a trustworthy fashion; and focusing on building long-term relationships" (Sherman 2009, 18).
3. Details on the conference proceedings are publicly available (https://businessandhumanrights.uconn.edu/events/stakeholder-engagement/, accessed May 3, 2018), and conference findings are presented in Leipziger (2018). The author's participant observation was covered under University of Connecticut Institutional Review Board Protocol #H16-294.
4. Cited in Sethi and Rovenpor (2016, 11), who note that the United Nations Conference on Trade and Development's *World Investment Report* 2011 draws on Litovsky, Rochlin, Zadek, and Levy's definition of MSIs (2007, 20).

5. For details, see the Worker-Driven Social Responsibility Network (https://wsr-network.org/what-is-wsr/, accessed March 7, 2018) and Gordon (2017, 7).
6. Rodríguez-Garavito notes the critique of MSIs that allow corporate-leaning NGOs to represent civil society instead of grassroots groups (i.e., the former include "new consulting firms and think tanks advising corporations and states on the issue of business and human rights"; see 2017a, 30). Elsewhere, I offer a three-part typology that arrays NGOs along a spectrum from less to more confrontational engagement with companies, based on variation in function (i.e., rule-making and enforcement vs. information-brokering vs. direction action and protest); see Hertel (2010, 178).
7. To illustrate managerial capture, Sethi and Ravenpor assess Fair Labor Association auditing reports of Hanes Brands' global operations from 2008 to 2014 and argue that the corrective action the company has taken in response to audit findings is incomplete (2016, 19).
8. Ambiguity creates the risk that actors "who want to avoid difficult interactions with critics and true interest representation" may entrench their control over the process of stakeholder engagement (Fransen and Kolk 2007, 679). Ambiguity also increases the likelihood that the "lowest common denominator" may emerge as the criterion for standards-setting and fulfillment (Fransen and Kolk 2007, 673).
9. Participant observation, covered under University of Connecticut Institutional Review Board Protocol #H16-294. See these observations reflected in the post-conference policy paper authored on behalf of the BHR Initiative by Deborah Leipziger (2018, 8, 11, 18, 20–21).
10. For competing views on the relationship between human rights and social justice, see Lettinga and Van Troost's collection (2015) and David Petrasek's corresponding synthesis therein (Petrasek 2015).
11. Sethi and Rovenpor note the relative inability of nation-states to create necessary regulatory oversight over MNCs [multinational corporations] along with unwillingness "to enforce their own labor laws. Furthermore, this situation

persists regardless of the nature of national government whether it is a democratically elected body or an authoritarian state" (2016, 9).
12. This reformist-to-radical continuum is discussed in relation to NGOs involved in labor rights advocacy more generally in Hertel (2010, 6).
13. Although the growing number of organizations in the MSI arena is too large to list here, key groups involved in research and training on MSIs include those discussed in Chapter 3: SHIFT (the NGO created by Ruggie and colleagues just days after the passage in 2011 of the United Nations Guiding Principles on Business and Human Rights, as a vehicle for implementing them); MSI Integrity (initially "incubated" at the Harvard Law Clinic from 2010 to 2012 and now an independent assessor of MSI practices); and numerous organizations that carry out supply chain auditing and compliance management, such as the Fair Labor Association and Social Accountability International.
14. The title of the appendix to the SHIFT report (2014a) characterizes this and other examples included as "Disclosure at the Leading Edge."
15. Interview VAG008053117 on May 31, 2017.

REFERENCES

Abouharb, M. Rodwan, and David Cingranelli. 2008. *Human Rights and Structural Adjustment*. New York/Cambridge, UK: Cambridge University Press.

Accord on Fire and Building Safety in Bangladesh. n.d. Accessed May 15, 2018. http://bangladeshaccord.org/.

Ackerly, Brooke. 2018. *Just Responsibility: A Human Rights Theory of Global Justice*. New York: Oxford University Press.

Action Aid. 2004. "Rethinking Participation: Questions for Civil Society About the Limitations of Participation in [Poverty Reduction Strategy Papers] PRSPs," ActionAid Uganda Discussion Paper. Accessed April 19, 2016. http://siteresources.worldbank.org/CSO/Resources/AA_Rethinking_Participation_by_Action_Aid.pdf.

Adler-Milstein, Sarah, and John M. Kline. 2017. *Sewing Hope: How One Factory Challenges the Apparel Industry's Sweatshops*. Oakland: University of California Press.

Albisa, Cathy. 2011. "Drawing Lines in the Sand: Building Economic and Social Rights Norms in the United States." In *Human Rights in the United States: Beyond Exceptionalism*, edited by Shareen Hertel and Kathryn Libal, 68–88. New York: Cambridge University Press.

Albisa, Cathy. 2018. "From the Women's March to the Poor People's Campaign, a Call for Economic Rights." *In These Times*, May

16, 2018. Accessed May 18, 2018. http://inthesetimes.com/working/entry/21147/poor-peoples-campaign-womens-march-human-rights-protest-human-rights.

Albisa, Cathy, and Ben Palmquist, with Brittany Scott, Sean Sellers, and Marigo Farr. 2018. *A New Social Contract: Collective Solutions Built By and For Communities*. New York: National Economic and Social Rights Initiative. Accessed May 26, 2018. https://www.nesri.org/news/2018/05/a-new-social-contract-transformative-solutions-built-by-and-for-communities.

Anner, Mark. 2011. *Solidarity Transformed: Labor Responses to Globalization and Crisis in Latin America*. Ithaca, NY: Cornell University/ILR Press.

Anner, Mark. 2017. "Monitoring Workers' Rights: The Limits of Voluntary Social Compliance Initiatives in Labor Repressive Regimes." *Global Policy* 8, no. 3: 56–65.

Anner, Mark. 2018. *Binding Power: The Sourcing Squeeze, Workers' Rights, and Building Safety in Bangladesh Since Rana Plaza*. Penn State Center for Global Workers' Rights Research Report, March 22, 2018. Accessed April 17, 2018. http://lser.la.psu.edu/gwr/documents/CGWR2017ResearchReportBindingPower.pdf.

Anner, Mark, Jennifer Bair, and Jeremy Blasi. 2013. "Toward Joint Liability in Global Supply Chains: Addressing the Root Causes of Labor Violations in International Subcontracting Networks." *Comparative Labor Law and Policy Journal* 35, no. 1: 1–44.

Anthony, Amanda K., and Ian M. Taplin. 2017. "Sustaining the Retail Pilgrimage: Developments of Fast Fashion and Authentic Identities." *Journal of Fashion, Style and Popular Culture* 4, no. 1: 33–50.

Asbed, Greg, and Sean Sellers. 2013. "The Fair Food Program: Comprehensive, Verifiable and Sustainable Change for Farmworkers." *University of Pennsylvania Journal of Law and Social Change* 16, no. 1: 39–48.

Backer, Larry Catá. 2012. "From Institutional Misalignments to Socially Sustainable Governance: The Guiding Principles for the Implementation of the United Nations Protect,

Respect and Remedy and the Construction of Inter-Systemic Global Governance." *Pacific McGeorge Global Business and Development Law Journal* 25, no. 1: 69–172.
Baer, Madeline. 2015. "From Water Wars to Water Rights: Implementing the Human Right to Water in Bolivia." *Journal of Human Rights* 14, no. 3: 353–376.
Bair, Jennifer, Marsha A. Dickson, and Doug Miller. 2014. "To Label or Not to Label: Is That the Question?" In *Workers' Rights and Labor Compliance in Global Supply Chains: Is a Social Label the Answer?*, edited by Jennifer Bair, Marsha A. Dickson, and Doug Miller, 3-22. London/New York: Routledge.
Barkemeyer, Ralf. 2009. "Beyond Compliance—Below Expectations? CSR in the Context of International Development." *Business Ethics: A European Review* 18, no. 3: 273–289.
Bauer, Joanne. 2011. "Business and Human Rights: A New Approach to Advancing Environmental Justice in the United States." In *Human Rights in the United States: Beyond Exceptionalism*, edited by Shareen Hertel and Kathryn Libal, 175–196. New York: Cambridge University Press.
Bauer, Joanne. 2016. "The Coalition of Immokalee Workers and the Campaign for Fair Food: The Evolution of a Business and Human Rights Campaign." In *Business and Human Rights: From Principles to Practice*, edited by Dorothee Baumann-Pauly and Justine Nolan, 175–178. New York: Routledge.
Baumann-Pauly, Dorothee, and Justine Nolan, eds. 2016. *Business and Human Rights: From Principles to Practice*. New York: Routledge.
Berliner, Daniel, Anne Regan Greenleaf, Milli Lake, Margaret Levi, and Jennifer Noveck. 2015. *Labor Standards in International Supply Chains: Aligning Rights and Incentives*. Cheltenham, UK: Edward Elgar.
Bird, Robert C., Daniel R. Cahoy, and Jamie Darin Prenkert, eds. 2014. *Law, Business and Human Rights: Bridging the Gap*. Cheltenham, UK/Northampton, MA: Edward Elgar.
Bissell, Susan L. 2005. "Earning and Learning: Tensions and Compatibility." In *Child Labor and Human Rights: Making Children Matter*, edited by Burns H. Weston, 377–399. Boulder, CO: Lynne Rienner Publishers.

BN Americas. 2018. Falconbridge Dominica summary. Accessed January 9, 2018. https://www.bnamericas.com/company-profile/en/falconbridge-dominicana-sa-falcondo.

Brea, Ramonina, Isis Duarte, and Mitchell Seligson. 2005. *Resultados de la IV Encuesta Nacional de Cultura Política y Democracia (Demos 2004)*. Santo Domingo: Pontificia Universidad Católica Madre y Maestra. Accessed February 14, 2017. http://www.vanderbilt.edu/lapop/dr/2004-cultura politica.pdf.

Bride, Michael. 2018. "Public Remarks by the Deputy Director of the Accord on Fire and Building Safety in Bangladesh." Delivered at the forum "Has Anything Changed Since Rana Plaza? Worker Rights and Safety, Five Years After the Devastation," New York, April 10, 2018.

Brooks, Ethel. 2007. *Unraveling the Garment Industry: Transnational Organizing and Women's Work*. Minneapolis: University of Minnesota Press.

Brown, Pins. 2007. "Principles That Make for Effective Governance of Multi-Stakeholder Initiatives." Prepared for the United Nations Special Representative of the Secretary General on Business and Human Rights Expert Workshop (November 6–7, 2007). Accessed February 28, 2018. http://business-humanrights.org/sites/default/files/media/bhr/files/Principles-for-effective-MSIs-6-7-Nov-2007.pdf.

Brysk, Alison. 2013. *Speaking Rights to Power: Constructing Political Will*. New York: Oxford University Press.

Buerger, Catherine, and Elizabeth Holzer. 2015. "How Does Community Participation Work? Human Rights and the Hidden Labour of Interstitial Elites in Ghana." *Journal of Human Rights Practice* 7, no. 1: 72–87.

BusinessWire. 2012. "Hanesbrands and Wake Forest Baptist Medical Center Deliver Healthcare Services to Children in the Dominican Republic." Accessed March 16, 2017. http://www.businesswire.com/news/home/20120510005927/en/HanesBrands-Wake-Forest-Baptist-Medical-Center-Deliver.

BusinessWire. 2015. "Hanes Brands Reports First Quarter 2015 Financial Results." Accessed July 24, 2017. https://www.businesswire.com/news/home/20150423006589/

en/HanesBrands-Reports-First-Quarter-2015-Financial-Results.
Büthe, Tim, and Walter Mattli. 2011. *The New Global Rulers: The Privatization of Regulation in the World Economy*. Princeton, NJ: Princeton University Press.
Callegaro, Mario. 2008. "Social Desirability." In *Encyclopedia of Survey Research Methods*, edited by Paul J. Lavrakas, 826. Thousand Oaks, CA: SAGE Publications, Inc.
Chapman, Audrey R. 1996. "A 'Violations Approach' for Monitoring the International Covenant on Economic, Social and Cultural Rights." *Human Rights Quarterly* 18, no. 1: 23–66.
Chapman, Audrey R. 2007. "The Status of Efforts to Monitor Economic, Social, and Cultural Rights." In *Economic Rights: Conceptual, Measurement, and Policy Issues*, edited by Shareen Hertel and Lanse Minkler, 143–164. New York: Cambridge University Press.
Chong, Daniel. 2010. *Freedom from Poverty: NGOs and Human Rights Praxis*. Philadelphia: University of Pennsylvania Press.
Clark, Ann Marie, Elisabeth J. Friedman, and Kathryn Hochstetler. 1998. "The Sovereign Limits of Global Civil Society: A Comparison of NGO Participation in UN World Conferences on the Environment, Human Rights, and Women." *World Politics* 51, no. 1: 1–35.
Collingsworth, Terry. 2003. "Separating Fact from Fiction in the Debate over the Application of the Alien Tort Claims Act to Violations of Fundamental Human Rights by Corporations." *University of San Francisco Law Review* 37, no. 3: 563–586.
Collins, Ben, Amelia Evans, Madeline Hung, and Suzanne Katzenstein. 2017. *The New Regulators? Assessing the Landscape of Multi-Stakeholder Initiatives*. San Francisco: MSI Integrity and the Duke Human Rights Center at the Kenan Institute for Ethics.
Columbia Law School Human Rights Clinic and International Human Rights Clinic at Harvard Law School. 2015. *Righting Wrongs? Barrick Gold's Remedy Mechanism for Sexual Violence in Papua New Guinea: Key Concerns and Lessons Learned*. New York and Boston: Columbia Law School and Harvard Law School.

Cooley, Alexander, and Jack Snyder, eds. 2015. *Ranking the World: Grading States as a Tool of Global Governance*. Cambridge, UK: Cambridge University Press.

Cornia, Giovanni Andrea, Richard Jolly, and Frances Stewart. 1987. *Adjustment with a Human Face*. Gloucestershire, UK: Clarendon Press, for United Nations Children's Fund (UNICEF).

Costanza, Jennifer N. 2015. "Indigenous Peoples' Right to Prior Consultation: Transforming Human Rights from the Grassroots in Guatemala." *Journal of Human Rights* 14, no. 2: 260–285.

Cutler, A. Claire. 1999. "Locating 'Authority' in the Global Political Economy." *International Studies Quarterly* 43, no. 1: 59–81.

Damico, Noelle, and Sean Sellers. 2018. *"Now the Fear Is Gone": Advancing Gender Justice Through Worker-Driven Social Responsibility*. Worker-Driven Social Responsibility Network. Accessed May 26, 2018. https://wsr-network.org/?s=now+the+fear+is+gone

Daniel, Caitlin, Joseph Wilde-Ramsing, Kris Genovese, and Virginia Sandjojo. 2015. *Remedy Remains Rare: An Analysis of 15 years of National Contact Point Cases and Their Contributions to Improve Access to Remedy for Victims of Corporate Misconduct*. Amsterdam, The Netherlands: OECD Watch.

de Felice, Damiano. 2014. "Measuring the Effectiveness of Grievance Mechanisms: Between Key Performance Indicators and Engagement with Affected Stakeholders." *Measuring Business and Human Rights Blog*. April 11, 2014. Accessed February 28, 2018. http://shiftproject.org/sites/default/files/May%202014%20Shift%20BLP%20Workshop%20Report%20Remediation.pdf.

Deitelhoff, Nicole, and Klaus Dieter Wolf. 2013. "Business and Human Rights: How Corporate Norm Violators Become Norm Entrepreneurs." In *The Persistent Power of Human Rights: From Commitment to Compliance*, edited by Thomas Risse, Stephen C. Ropp, and Kathryn Sikkink, 222–238. New York/Cambridge, UK: Cambridge University Press.

Deutsche Gesellschaft fur Internationale Zusammenarbeit/GIZ. 2011. "Stakeholder Dialogues Manual." Eschborn, Germany: GIZ. Accessed February 28, 2018. http://www.mspguide.org/sites/default/files/resource/giz_stakeholder_dialogues_kuenkel.pdf.

Deva, Surya. 2012. *Regulating Corporate Human Rights Violations: Humanizing Business*. London/New York: Routledge.
Deva, Surya, and David Bilchitz, eds. 2013. *Human Rights Obligations of Business: Beyond the Corporate Responsibility to Respect?* New York: Cambridge University Press.
Dickson, Marsha A., Suzanne Loker, and Molly Eckman. 2009. *Social Responsibility in the Global Apparel Industry*. New York: Fairchild Books.
Dijkstra, Geske. 2011. "The PRSP Approach and the Illusion of Improved Aid Effectiveness: Lessons from Bolivia, Honduras and Nicaragua." *Development Policy Review* 1, no. 29: 111–133.
Donaghey, Jimmy, and Juliane Reinecke. 2018. "When Industrial Democracy Meets Corporate Social Accountability—A Comparison of the Bangladesh Accord and Alliance as Responses to the Rana Plaza Disaster," *British Journal of Industrial Relations* 56, no. 1: 14–42.
Eckerman, Ingrid. 2005. *The Bhopal Saga—Causes and Consequences of the World's Largest Industrial Disaster*. Telangana, India: Universities Press.
Elkins, Julie, and Shareen Hertel. 2011. "Sweatshirts and Sweatshops: Labor Rights, Student Activism, and the Challenges of Collegiate Apparel Manufacturing." In *Human Rights in Our Own Backyard: Injustice and Resistance in the United States*, edited by William T. Armaline, Davita Glasberg, and Bandana Purkayastha, 9–21. Philadelphia: University of Pennsylvania Press.
Fine, Janice. 2017. "Enforcing Labor Standards in Partnership with Civil Society: Can Co-Enforcement Succeed Where the State Alone Has Failed?" *Politics and Society* 45, no. 3: 359–388.
Fine, Janice, and Jennifer Gordon. 2010. "Strengthening Labor Standards Enforcement Through Partnership with Workers' Organizations." *Politics and Society* 38, no. 4: 552–585.
Fine, Janice, and Allison J. Petrozziello. 2017. "Haitian Migrant Workers in the Dominican Republic: Organizing at the Intersection of Informality and Illegality." In *Informal Workers and Collective Action: A Global Perspective*, edited by Adrienne E. Eaton, Susan J. Schurman, and Martha A. Chen, 71–95. Ithaca, NY: Cornell University Press.

Foucault, Michel. 1980. *Language, Counter-Memory, and Practice: Selected Essays and Interviews*, translated and edited by Donald F. Bouchard. Ithaca, NY: Cornell University Press.

Fransen, Luc W., and Brian Burgoon. 2017. "Introduction to the Special Issue: Public and Private Labor Standards Policy in the Global Economy." *Global Policy* 8, no. 3: 5–14.

Fransen, Luc W., and Ans Kolk. 2007. "Global Rule-Setting for Business: A Critical Analysis of Multi-Stakeholder Standards." *Organization* 14, no. 5: 667–684.

Fulton, Taylor, Jinhwa Ha, Michael Karimian, Eva Lerner, Annalisa Castillo Meier, and Isabelle Plessis. 2015. *What Is Remedy for Corporate Human Rights Abuses? Listening to Community Voices: A Field Report*. New York: Columbia University School of International & Public Affairs.

Fung, Archon, and Erik Olin Wright. 2003. "Countervailing Power in Empowered Participatory Governance." In *Deepening Democracy: Institutional Innovations in Empowered Participatory Governance*, edited by Archon Fun and Erik O. Wright, 259–290. London: Verso.

Gereffi, Gary, John Humphrey, and Timothy Sturgeon. 2005. "The Governance of Global Value Chains." *Review of International Political Economy* 12, no. 1: 78–104.

Godoy, Angelina Snodgrass. 2018. "Making Meaning of Violence: Human Rights and Historical Memory in the Conflict in El Salvador." *Journal of Human Rights* 17, no. 3: 367–379.

Goodhart, Michael. 2018. "Constructing Dignity: Human Rights as a Praxis of Egalitarian Freedom." *Journal of Human Rights* 17, no. 4. Published ahead of print, April 4, 2018. https://doi.org/10.1080/14754835.2018.1450738.

Gordon, Jennifer. 2017. The Problem with Corporate Social Responsibility. Remarks adapted from a public talk delivered June 11, 2014, at the Open Society Foundation, New York City. Accessed March 13, 2018. https://wsr-network.org/resource/the-problem-with-corporate-social-responsibility/.

Gray, Margaret, and Shareen Hertel. 2009. "Immigrant Farmworker Advocacy: The Dynamics of Organizing." *Polity* 41, no. 4: 409–435.

Hanes Brands. 2017. *Annual Report*. Winston-Salem, NC: Hanes Brands. Accessed July 24, 2017. http://ir.hanesbrands.com/phoenix.zhtml?c=200600&p=irol-reportsannual.

Hanes Corporation. 2017. "HBI-Owned Facilities." Accessed May 30, 2018. https://hanesforgood.com/content/uploads/2017/04/2017-HBI-Self-Owned.pdf.

Hayner, Priscilla. 2010. *Unspeakable Truths: Transitional Justice and the Challenge of Truth Commissions*. New York: Routledge.

Helwege, Ann. 2015. "Challenges with Resolving Mining Conflicts in Latin America." *Extractive Industries and Society* 2, no. 1: 73–84.

Hertel, Shareen. 2005. "What Was All the Shouting About? Strategic Bargaining and Protest at the WTO Third Ministerial (Seattle, Washington USA—1999)." *Human Rights Review* 6, no. 3: 102–118.

Hertel, Shareen. 2006. *Unexpected Power: Conflict and Change Among Transnational Activists*. Ithaca, NY: Cornell University Press.

Hertel, Shareen. 2010. "The Paradox of Partnership: Assessing New Forms of NGO Advocacy on Labor Rights." *Ethics & International Affairs* 24, no. 2: 171–189.

Hertel, Shareen, and Lanse P. Minkler, eds. 2007. *Economic Rights: Conceptual, Measurement and Policy Issues*. New York: Cambridge University Press.

Hertel, Shareen, Lyle Scruggs, and C. Patrick Heidkamp. 2009. "Human Rights and Public Opinion: From Attitudes to Action." *Political Science Quarterly* 124, no. 3: 445–461.

Hirschman, Albert O. 1970. *Exit, Voice and Loyalty*. Cambridge, MA: Harvard University Press.

Hopgood, Stephen. 2013. *The Endtimes of Human Rights*. Ithaca, NY: Cornell University Press.

Human Rights Watch. 2001. "Qatar: Inappropriate Venue for Next WTO Meeting." New York: Human Rights Watch. Accessed March 8, 2017. https://www.hrw.org/news/2001/01/19/qatar-inappropriate-venue-next-wto-meeting.

Human Rights Watch. 2011. *Gold's Costly Dividend: Human Rights Impacts of New Guinea's Porgera Gold Mine*. New York: Human Rights Watch. Accessed October 11, 2017. https://www.

hrw.org/report/2011/02/01/golds-costly-dividend/human-rights-impacts-papua-new-guineas-porgera-gold-mine.

International Alert and Engineers Against Poverty. 2006. "Conflict Sensitive Business Practice: Engineering Contractors and Their Clients." London: International Alert and EAP. Accessed February 28, 2018. http://www.international-alert.org/publications/conflict-sensitive-business-practice-engineering-contractors-and-their-clients.

International Petroleum Industry Environmental Conservation Association/IPIECA. 2012. "Operational Level Grievance Mechanisms: IPIECA Good Practice Survey." London: IPIECA. Accessed February 28, 2018. http://accessfacility.org/sites/default/files/IPIECA%20-%20Operational-level%20Grievance%20Mechanisms%3B%20Good%20Practice%20Survey.pdf.

Jerbi, Scott. 2009. "Business and Human Rights at the UN: What Might Happen Next." *Human Rights Quarterly* 31, no. 2: 299–320.

Kaeb, Caroline, and David Scheffer. 2011. "The Five Levels of CSR Compliance: The Resiliency of Corporate Liability Under the Alien Tort Statute and the Case for a Counterattack Strategy in Compliance Theory." *Berkeley Journal of International Law* 29, no. 1: 334–397. Republished in *Human Rights Obligations of Non-State Actors*, 2nd ed., edited by Andrew Clapham, 655–718. New York: Oxford University Press, 2013.

Kaplan, Rami. 2015. "Who Has Been Regulating Whom, Business or Society? The Mid-20th Century Institutionalization of 'Corporate Responsibility' in the USA." *Socio-Economic Review* 13, no. 1: 125–155.

Kaufman, Jonathan, and Katherine McDonnell. 2015. "Community-Driven Operational Grievance Mechanisms." *Business and Human Rights Journal* 1, no. 1: 127–132.

Keck, Margaret E., and Kathryn Sikkink. 1998. *Activists Beyond Borders: Advocacy Networks in International Politics*. Ithaca, NY: Cornell University Press.

Kinderman, Daniel. 2012. "'Free Us Up So We Can Be Responsible!' The Co-evolution of Corporate Social Responsibility and

Neo-liberalism in the UK, 1977–2010." *Socio-Economic Review* 10, no. 1: 29–57.

Kinderman, Daniel. 2015. "Explaining the Rise of National Corporate Social Responsibility: The Role of Global Frameworks, World Culture, and Corporate Interests." In *Corporate Social Responsibility in a Globalizing World*, edited by Kiyoteru Tsutsui and Alwyn Lim, 107–146. New York: Cambridge University Press.

Kingstone, Peter, ed. 2011. *The Political Economy of Latin America: Reflections on Neoliberalism and Development*: New York: Routledge.

Kline, John M. 2010. *Alta Gracia: Branding Decent Work Conditions—Will College Loyalty Embrace "Living Wage" Sweatshirts?* Research report prepared for the Kalmanovitz Initiative for Labor and the Working Poor. Washington, DC: Georgetown University.

Kline, John M., and Edward Soule. 2011. *Research Progress Report—Alta Gracia: Work with a Salario Digno*. Report prepared for the Reflective Engagement Initiative. Washington, DC: Georgetown University.

Kline, John M., and Edward Soule. 2014. *Research Results Report—Alta Gracia: Four Years and Counting*. Report prepared for the Reflective Engagement Initiative. Washington, DC: Georgetown University.

Knuckey, Sarah, and Eleanor Jenkin. 2015. "Company-Created Remedy Mechanisms for Serious Human Rights Abuses: A Promising New Frontier for the Right to Remedy?" *International Journal of Human Rights* 19, no. 6: 801–827.

Landefeld, John C., Katherine B. Burmaster, David H. Rehkopf, S. Leonard Syme, Maureen Lahiff, Sarah Adler-Milstein, and Lia C. H. Fernald. 2014. "The Association Between a Living Wage and Subjective Social Status and Self-Rated Health: A Quasi-experimental Study in the Dominican Republic." *Social Science and Medicine* 121: 91–97.

Lazarus, Joel. 2008. "Participation in Poverty Reduction Strategy Papers: Reviewing the Past, Assessing the Present and Predicting the Future." *Third World Quarterly* 29, no. 6: 1205–1221.

LeBaron, Genevieve, Jane Lister, and Peter Dauvergne. 2017. "Governing Global Supply Chain Sustainability through the Ethical Audit Regime." *Globalizations* 14, no. 6: 958–975.

Leipziger, Deborah. 2018. *Protecting Rights at the End of the Line: Stakeholder Engagement in Light Manufacturing*. Storrs, CT: University of Connecticut Business and Human Rights Initiative. Accessed May 18, 2018. https://businessandhumanrights.uconn.edu/stakeholder-engagement/.

Lettinga, Doutje, and Lars Van Troost, eds. 2015. *Can Human Rights Bring Social Justice? Twelve Essays*. Amsterdam, The Netherlands: Amnesty International Netherlands.

Litovsky, Alejandro, Steven Rochlin, Simon Zadek, and Brian Levy. 2007. *Investing in Standards for Sustainable Development: The Role of International Development Agencies in Supporting Collaborative Standards Initiatives*. London: AccountAbility.

Locke, Richard M. 2013. *The Promise and Limits of Private Power: Promoting Labor Standards in a Global Economy*. New York/Cambridge: Cambridge University Press.

Lopez, Linette. 2017. "One Company in the Western Hemisphere Has Thrown Politics Completely Off-Kilter." *Business Insider*, May 30, 2017. Accessed June 29, 2017. http://www.businessinsider.com/what-is-the-odebrecht-corruption-scandal-2017-5.

Malone, Christopher, Meghana Nayak, Matthew Bolton, and Emily Welty, eds. 2013. *Occupying Political Science: The Occupy Wall Street Movement from New York to the World*. Basingstoke, UK: Palgrave MacMillan.

Marx, Axel, Jane Wouters, Glenn Rayp, and Laura Beke, eds. 2015. *Global Governance of Labour Rights: Assessing the Effectiveness of Transnational Public and Private Policy Initiatives*. Cheltenham, UK/Northhampton, MA: Edward Elgar Publishing.

McLagan, Meg. 2003. "Principles, Publicity, and Politics: Notes on Human Rights Media." *American Anthropologist* 105, no. 3: 605–612.

Melish, Tara. 2017. "Putting 'Human Rights' Back into the UN Guiding Principles on Business and Human Rights: Shifting

Frames and Embedding Participation Rights." In *Business and Human Rights: Beyond the End of the Beginning*, edited by César Rodríguez-Garavito, 76–96. New York/Cambridge, UK: Cambridge University Press.

Merriam-Webster Dictionary. 2017. Springfield, MA: Merriam-Webster, Inc. Online version. Accessed February 28, 2016. https://www.merriam-webster.com/dictionary/dictionary.

Merry, Sally Engle, Kevin E. Davis, and Benedict Kingsbury, eds. 2015. *The Quiet Power of Indicators: Measuring Governance, Corruption and Rule of Law*. New York/Cambridge, UK: Cambridge University Press.

Meyers, Diana Tietjens. 2016. *Victims Stories and the Advancement of Human Rights*. New York: Oxford University Press.

Meyersfeld, Bonita. 2017. "Committing the Crime of Poverty: The Next Phase of the Business and Human Rights Debate." In *Business and Human Rights: Beyond the End of the Beginning*, edited by César Rodríguez-Garavito, 173–185. New York: Cambridge University Press.

Mill, John S. 1843. *A System of Logic*. London: John W. Parker.

Moleres, Fernando. 2000. *Stolen Childhood*. Geneva, Switzerland: International Labour Organization.

Morgan, Jana, Rosairo Espinal, and Mitchell Seligson. 2006. *The Political Culture of Democracy in Dominican Republic: 2006*. Nashville, TN: Vanderbilt University Latin American Public Opinion Project (LAPOP). Accessed February 14, 2017. http://www.vanderbilt.edu/lapop/dr/2006-political culture.pdf.

Mosley, Layna, and Saika Uno. 2007. "Racing to the Bottom or Climbing to the Top? Economic Globalization and Collective Labor Rights." *Comparative Political Studies* 40, no. 8: 923–948.

Moyn, Samuel. 2010. *The Last Utopia: Human Rights in History*. Cambridge, MA: The Belknap Press of Harvard University Press.

Mwangi, Wagaki, Lothar Rieth, and Hans Peter Schmitz. 2013. "Encouraging Greater Compliance: Local Networks and the United Nations Global Compact." In *The Persistent Power of Human Rights: From Commitment to Compliance*, edited by

Thomas Risse, Stephen C. Ropp, and Kathryn Sikkink, 203–221. New York: Cambridge University Press.

Nadvi, Khalid. 2004. "Globalisation and Poverty: How Can Global Value Chain Research Inform the Policy Debate?" *IDS Bulletin* 35, no. 1: 20–30.

Narayan, Deepa, ed. 2005. *Measuring Empowerment: Cross-Disciplinary Perspectives*. Washington, DC: The World Bank.

Nelson, Paul J., and Ellen Dorsey. 2008. *New Rights Advocacy: Changing Strategies of Development and Human Rights NGOs*. Washington, DC: Georgetown University Press.

Nolan, Hamilton. 2018. "How Power Is Built from the Bottom: CTUL, Minneapolis' Worker Center Movement." *Splinter News*, January 17, 2018. Accessed March 14, 2018. https://www.nesri.org/print/3210.

O'Connor, Casey, and Sarah Labowitz. 2017. *Putting the "S" in ESG: Measuring Human Rights Performance for Investors*. New York: New York University Stern Center for Business and Human Rights.

O'Rourke, Dara. 2003. "Outsourcing Regulation: Analyzing Nongovernmental Systems of Labor Standards and Monitoring." *Policy Studies Journal* 31, no. 1: 1–29.

Pattipatti, Krishna. 2017. "Inference and Optimization Applications to Complex Systems: Board of Trustees Distinguished Professor Lecture." Remarks at the University of Connecticut School of Engineering, Storrs, CT, September 29, 2017. Accessed January 16, 2018. https://www.youtube.com/watch?v=Wsjx3p7emw4.

Petrasek, David. 2015. "Human Rights and Social Justice—A False Dichotomy." In *Can Human Rights Bring Social Justice? Twelve Essays*, edited by Doutje Lettinga and Lars Van Troost, 89–99. Amsterdam, The Netherlands: Amnesty International Netherlands.

Pineda, Jorge. 2017. "Dominican Republic Arrests Officials in Odebrecht Bribery Probe." *Reuters*, May 29, 2017. Accessed May 29, 2017. https://www.reuters.com/article/us-dominican-corruption-idUSKBN18P1S7.

Pogge, Thomas. 2008. *World Poverty and Human Rights*. Cambridge, UK: Polity Press.

Pruce, Joel R., and Alexandra Cosima Budabin. 2016. "Beyond Naming and Shaming: New Modalities of Information Politics in Human Rights." *Journal of Human Rights* 15, no. 3: 408–425.

Ramasastry, Anita. 2015. "Corporate Social Responsibility Versus Business and Human Rights: Bridging the Gap Between Responsibility and Accountability." *Journal of Human Rights* 14, no. 2: 237–259.

Reese, Caroline. 2011. "Piloting Principles for Effective Company–Stakeholder Grievance Mechanisms: A Report of Lessons Learned." Cambridge, MA: Corporate Social Responsibility Initiative, Harvard Kennedy School. Accessed February 28, 2018. http://www.hks.harvard.edu/m-rcbg/CSRI/publications/report_46_GM_pilots.pdf.

Reinecke, Juliane, and Jimmy Donaghey. 2015. "The 'Accord for Fire and Building Safety in Bangladesh' in Response to the Rana Plaza Disaster." In *Global Governance of Labour Rights: Assessing the Effectiveness of Transnational Public and Private Policy Initiatives*, edited by Axel Marx, Jane Wouters, Glenn Rayp, and Laura Beke, 257–277. Cheltenham, UK/ Northhampton, MA: Edward Elgar Publishing.

República Dominicana Sistema Interactiva de Consulta Censo/SICEN. 2010. Accessed January 4, 2017. http://sicen.one.gob.do/.

Risse, Thomas, Stephen C. Ropp, and Kathryn Sikkink, eds. 1999. *The Power of Human Rights: International Norms and Domestic Change*. New York: Cambridge University Press.

Risse, Thomas, Stephen C. Ropp, and Kathryn Sikkink, eds. 2013. *The Persistent Power of Human Rights: From Commitment to Compliance*. New York: Cambridge University Press.

Ristovska, Sandra. 2016. "The Rise of Eyewitness Video and Its Implications for Human Rights: Conceptual and Methodological Approaches." *Journal of Human Rights* 15, no. 3: 347–360.

Rivoli, Petra. 2015. *The Travels of a T-Shirt in the Global Economy: An Economist Examines the Markets, Power, and Politics of World Trade*, 2nd ed. Hoboken, NJ: Wiley & Sons.

Rocha, Lola, Eduardo Bustelo, Ernesto López, and Luis Zúñiga, eds. 1989. "Women, Economic Crisis and Adjustment

Policies: Interpretation and Initial Assessment." In *The Invisible Adjustment: Poor Women and the Economic Crisis*, 9–27. Santiago, Chile: UNICEF, The Americas and the Caribbean Regional Office.

Rodney, Walter. 1972. *How Europe Underdeveloped Africa*. London, UK: Bogle L'Overture Publications.

Rodriguez, Amaury. 2017. "A Green Tide Engulfs the DR." *NACLA News*, June 16, 2017. Accessed February 20, 2018. https://nacla.org/news/2017/06/19/%E2%80%9Cgreen-tide%E2%80%9D-engulfs-dr.

Rodríguez-Garavito, César, ed. 2017a. *Business and Human Rights: Beyond the End of the Beginning*. New York: Cambridge University Press.

Rodríguez-Garavito, César. 2017b. "Introduction: A Dialogue Across Divides in the Business and Human Rights Field." In *Business and Human Rights: Beyond the End of the Beginning*, edited by César Rodríguez-Garavito, 1–8. New York/Cambridge, UK: Cambridge University Press.

Rottenburg, Richard, Sally E. Merry, Sung-Joon Park, and Johanna Mugler, eds. 2015. *The World of Indicators: The Making of Governmental Knowledge through Quantification*. New York/Cambridge, UK: Cambridge University Press.

Ruggie, John. 2013 *Just Business: Multinational Corporations and Human Rights*. New York: WW Norton and Co., Inc.

Saiz, Ignacio. 2018. "Human Rights Must Rise to the Challenge of Growing in Equality, Not Retreat in Defeat." Center for Economic and Social Rights. Accessed May 18, 2018. http://www.cesr.org/human-rights-must-rise-challenge-growing-inequality-not-retreat-defeat.

Salgado, Soli. 2016. "Mining Our Mountain: People in the Dominican Republic Tell a Canadian Company 'No.'" *Global Sisters Report*, September 22, 2016. Accessed January 9, 2018. http://globalsistersreport.org/news/environment/mining-our-mountain-people-dominican-republic-tell-canadian-company-no-42366.

Schrank, Andrew. 2011. "Co-producing Workplace Transformation: The Dominican Republic in Comparative Perspective." *Socio-Economic Review* 9, no. 3: 419–445.

Schrank, Andrew. 2013. "From Disguised Protectionism to Rewarding Regulation: The Impact of Trade-Related Labor Standards in the Dominican Republic." *Regulation and Governance* 7, no. 3: 299–320.

Scruggs, Lyle, Shareen Hertel, Samuel Best, and Christopher Jeffords. 2011. "Information, Choice and Political Consumption: Human Rights in the Checkout Lane." *Human Rights Quarterly* 33, no. 4: 1092–1121.

Sen, Amartya. 1999. *Development as Freedom*. Oxford, UK: Oxford University Press.

Sethi, Prakash, and Janet I. Rovenpor. 2016. "The Role of NGOs in Ameliorating Sweatshop-Like Conditions in the Global Supply Chain: The Case of Fair Labor Association (FLA) and Social Accountability International (SAI)." *Business and Society Review* 5, no. 1: 5–36.

Sherman, John. 2009. *Embedding Rights-Compatible Grievance Processes for External Stakeholders Within Business Culture*. Cambridge, MA: Corporate Social Responsibility Initiative, Harvard Kennedy School. Accessed February 28, 2018. https://www.shiftproject.org/resources/publications/embedding-rights-compatible-grievance-processes-external-stakeholders-within-business-culture/.

SHIFT. 2014a. *Evidence of Corporate Disclosure relevant to the UN Guiding Principles on Business and Human Rights*. New York: SHIFT.

SHIFT. 2014b. *Remediation, Grievance Mechanisms and the Corporate Responsibility to Respect Human Rights*. SHIFT Workshop Report No. 5. Accessed November 1, 2015. http://shiftproject.org/sites/default/files/May%202014%20Shift%20BLP%20Workshop%20Report%20Remediation.pdf.

Skinner, Gwynne, Robert McCorquodale, Olivier De Schutter, and Andie Lambe, for the International Corporate Accountability Roundtable, CORE, and the European Coalition for Corporate Justice. 2013. *The Third Pillar: Access to Judicial Remedies for Human Rights Violations by Transnational Business*. Accessed November 1, 2015. http://icar.ngo/wp-content/uploads/2013/02/The-Third-Pillar-Access-to-Judicial-Remedies-for-Human-Rights-Violation-by-Transnational-Business.pdf.

Soundararajan, Vivek, and Jill A. Brown. 2016. "Voluntary Governance Mechanisms in Global Supply Chains: Beyond CSR to a Stakeholder Utility Perspective." *Journal of Business Ethics* 134, no. 1: 83–102.

Soundararajan, Vivek, Jill A. Brown, and Andrew C. Wicks. 2016. "Value Creation Through Multistakeholder Initiatives: An Instrumental Perspective." Paper presented at the Academy of Management Conference, Anaheim, California.

The Economist. 2013. "Mining in the Dominican Republic, Sickness and Wealth." September 21, 2013. Accessed January 9, 2018. https://www.economist.com/news/americas/21586560-shiny-new-mine-rusty-pollution-problems-sickness-and-wealth.

Transparency International. 2017. *Corruption Perceptions Index 2017.* Accessed March 1, 2018. https://www.transparency.org/news/feature/corruption_perceptions_index_2017.

United Nations Conference on Trade and Development. 2011. *World Investment Report 2011.* New York: UNCTAD. Accessed March 15, 2018. http://unctad.org/en/docs/wir2011_embargoed_en.pdf.

United Nations Development Programme. 2015. *Human Development Report 2015*, "Dominican Republic" Briefing Note. Accessed January 4, 2017. http://hdr.undp.org/sites/all/themes/hdr_theme/country-notes/DOM.pdf.

United Nations Development Programme. 2017. *El embarazo en adolescentes: un desafío multidimensional para generar oportunidades en el ciclo de vida.* Santo Domingo: PNUD República Dominicana. Accessed February 22, 2018. http://hdr.undp.org/sites/default/files/reports/2831/pnud_do_indh2017web.pdf.

United Nations General Assembly. 2013. Report of the Special Rapporteur on Extreme Poverty and Human Rights, Magdalena Sepúlveda Carmona, to the Human Rights Council, 23rd Session, Agenda Item 3, A/HRC/23/26.

United Nations Global Compact. 2011. *UN Global Compact Local Network Report 2011.* New York: UN Global Compact. Accessed December 21, 2018. https://www.unglobalcompact.org/docs/networks_around_world_doc/Annual_Report_2011/Annual_Local_Network_Report_2011.pdf

United Nations Global Compact. 2015. *Impact: Transforming Business, Changing the World*. New York and Oslo, Norway: DNV GL.

United Nations Human Rights Council. 2011. "Guiding Principles on Business and Human Rights: Implementing the United Nations 'Protect, Respect and Remedy' Framework." A/HRC/17/31 (March 21, 2011). Accessed February 28, 2018. http://www.ohchr.org/Documents/Publications/GuidingPrinciplesBusinessHR_EN.pdf.

United States Department of State, Overseas Security Advisory Council. 2014. *Dominican Republic 2014 Crime and Safety Report*. Washington, DC: US Department of State. Accessed June 22, 2017. https://www.osac.gov/Pages/ContentReportDetails.aspx?cid=16417.

VanEvera, Stephen. 1997. *Guide to Methods for Students of Political Science*. Ithaca, NY: Cornell University Press.

Vogel, David. 2006. *The Market for Virtue: The Potential and Limits of Corporate Social Responsibility*. Washington, DC: Brookings Institution Press.

Wettstein, Florian. 2012. "CSR and the Debate on Business and Human Rights." *Business Ethics Quarterly* 22, no. 4: 739–770.

Wilson, Emma, and Emma Blackmore, eds. 2013. *Dispute or Dialogue? Community Perspectives on Company-Led Grievance Mechanisms*. London: International Institute for Environment and Development. Accessed February 28, 2018. http://pubs.iied.org/pdfs/16529IIED.pdf.

Winston, Morton E., and John C. Pollock. 2016. "Introduction: Human Rights in the News—Balancing New Media Participation with the Authority of Journalism and Human Rights Professionalism." *Journal of Human Rights* 15, no. 3: 307–313.

Worker-Driven Responsibility (WSR) Network. n.d. Accessed May 18, 2018. https://wsr-network.org/.

Workers' Rights Consortium. 2007. WRC Assessment re: TOS Dominicana (Dominican Republic): Findings and Recommendations of June 6, 2007. Washington, DC: WRC. Accessed March 16, 2017. http://digitalcommons.ilr.cornell.edu/cgi/viewcontent.cgi?article=1382&context=globaldocs.

Workers' Rights Consortium. 2008. *WRC Factory Investigation: TOS Dominica, Updated August 22.* Accessed April 18, 2018. http://accessfacility.org/sites/default/files/Comments%20on%20Hanesbrands%20Report%20re%20TOS%20Dominicana%20-%203-12-08.pdf.

Workers' Rights Consortium. 2010. *Living Wage Analysis for the Dominican Republic.* Washington, DC: WRC. Accessed April 17, 2018. http://www.workersrights.org/linkeddocs/WRC%20Living%20Wage%20Analysis%20for%20the%20Dominican%20Republic.pdf.

Workers' Rights Consortium. 2013. *Global Wage Trends for Apparel Workers, 2001–2011.* Washington, DC: WRC. Accessed February 28, 2018. https://www.americanprogress.org/issues/economy/reports/2013/07/11/69255/global-wage-trends-for-apparel-workers-2001-2011/.

World Bank, Operations Evaluation Department. 2004. *The Poverty Reduction Strategy Initiative: An Independent Evaluation of the World Bank's Support Through 2003.* Washington, DC: World Bank OED. Accessed March 8, 2017. http://lnweb90.worldbank.org/oed/oeddoclib.nsf/DocUNIDViewForJavaSearch/6B5669F816A60AAF85256EC1006346AC/$file/PRSP_Evaluation.pdf.

Zack, Suzanne. 2015. "Papers and Media Archive of Filmmaker and Human Rights Advocate U. Roberto Romano Given to UConn's Archives and Special Collections." *UConn Libraries News.* October 6, 2015. Accessed March 8, 2017. http://blogs.lib.uconn.edu/news/author/szack/#.VsyRcvkrKUk.

Zadek, Simon. 2004. "The Path to Corporate Responsibility." *Harvard Business Review* 48, no. 12: 125–132.

INDEX

Tables are indicated by f following the page number.

Accord on Fire and Building
 Safety in Bangladesh,
 141–44, 155–56, 160–61
Ackerly, Brooke, 139–42, 161–62
Adjustment with a Human Face
 (UNICEF), 17
Adler-Milstein, Sarah, 62, 63–64,
 72–73, 82, 129
agricultural industries
 labor rights in, 154–56
 multi-stakeholder initiatives
 in, 47–48
Alien Tort Claims Act (US),
 19–20, 23–24
Alta Gracia Apparel in Villa
 Altagracia
 collective action challenges
 at, 93
 community impact of, 55, 64
 community perceptions of,
 82, 88

 living wage paid by, 52, 53,
 59–60, 62, 63–64, 72–73,
 94–95, 143–44
 WSR business model for,
 52, 59–60, 95, 146–47,
 155–56, 163–64
Americano Nickel Ltd., 99–100
Anner, Mark, 44–45, 143–44
Asbed, Greg, 155–56
automotive sector, 47–48

Bair, Jennifer, 143–44
Bangladesh
 child labor in, 159–61
 factory safety initiatives
 in, 141–44
 light manufacturing
 stakeholder consultation
 in, 42–43
Barrick Gold, 99–100, 112, 126
Bauer, Joanne, 20

Bhopal disaster (1984), 19–20
BHR. *See* business and human rights framework
BHRRC. *See* Business and Human Rights Resource Centre
Bissell, Susan, 159–60
Blasi, Jeremy, 143–44
Bonao (Dominican Republic), 98–132
  challenges for residents in, 102–5
  community attitudes on nature of remedy in, 125–29, 147
  constraints on stakeholder dialogue in, 118–21
  corruption in, 120–21, 127
  crime in, 104–5
  data collection in, 56–57
  demographics of, 52–53
  Dos Rios free-trade zone in, 98–99, 108, 112, 127
  electricity outages in, 75–76, 103–4, 124–25
  food insecurity in, 102
  industrial economy in, 105–7
  infrastructure challenges in, 75–76, 103–4, 124–25
  knowledge of stakeholder dialogue in, 121–25
  labor market in, 107–10
  lessons learned, 129–31
  local context for, 57–60
  policy implications, 131–32
  power-based analysis in, 154
  receptivity to stakeholder dialogue in, 121–25
  responsibility for community well-being in, 113–18
  stakeholder perceptions in, 98–102
  subjective socioeconomic status in, 110–13, 129
  worker training in, 107–10
Boston Consulting Group, 33–34
Brazil
  extractive industries stakeholder consultation in, 42–43
  light manufacturing stakeholder consultation in, 42–43
Brent Spar oil platform, 19
Brown, Jill A., 31–33, 34–35
business and human rights (BHR) framework
  corporate involvement in, 6–7
  emergence of, 1–2
  entitlement vehicle phase and, 30
  limitations of, 2
  mapping trends in stakeholder consultation, 37–39
  social legitimacy of, 6–7
Business and Human Rights Resource Centre (BHRRC)
  API for data analysis, 38–40, 165–66
  as data source, 12, 37–39

Chapman, Audrey, 10
child labor, 159–61
China, light manufacturing stakeholder consultation in, 42–43
citizenship rights, 133, 145
civil society organizations. *See also specific organizations*
  BHR and, 1–2
  participation rates in, 33–34
Clean Clothes Campaign, 23–24
Coalition of Immokalee Workers, 155–56
collective action, 92–93
Colombia
  extractive industries stakeholder consultation in, 42–43

INDEX | **213**

light manufacturing
 stakeholder consultation
 in, 42–43
community well-being,
 responsibility for
 in Bonao, 113–18
 community perceptions of, 53
 in Villa Altagracia, 77–80, 91
 commuting costs, 68–69, 70–71
 complaints mechanism, 40,
 41–42, 48–49
conditional cash transfers, 72
Congo. See Democratic Republic
 of Congo
consumer goods industries, 47–48
Convention Against
 Corruption, 144
Convention on Access to
 Information, Public
 Participation in Decision-
 Making and Access to
 Justice in Environmental
 Matters, 144
Convention on the Rights of
 Persons with Disabilities, 144
Cooley, Alexander, 17–18
corporate disclosure policies,
 47, 48–49
corruption, 74–75, 78–79,
 120–21, 127, 131–32, 144
crime, 73–74, 75–76, 104–5
damage control phase
 of stakeholder
 consultation, 16–21

Davis, Kevin E., 17–18
Deitelhoff, Nicole, 7
Democratic Republic of Congo
 extractive industries
 stakeholder consultation
 in, 42–43
 light manufacturing stakeholder
 consultation in, 42–43

discretionary philanthropy,
 87, 91–92
distributive justice, 32, 60
Dominican Republic. *See also*
 Bonao; Villa Altagracia
 corruption in, 131–32
 data collection in, 56–57
 demographics of, 66–67
 development strategy in, 66
 extractive industries in, 100
 gender inequality in, 67
 living wage in, 63–64
 local context for, 57–60
 *Ministerio de Economía,
  Planificación y Desarrollo*
  (MEPyD), 57, 117
 Ministerio de la Mujer (Ministry
  of Women), 57
 Ministry of Labor in, 107–8
 political context in, 65–66
 poverty rate in, 65–66
 stakeholder consultation case
  studies in, 51–55
Donaghey, Jimmy, 143
Dos Rios free-trade zone (Bonao),
 98–99, 108, 112, 127

economic rights
 in light manufacturing
  communities, 4–5, 53–54
 policymaking for, 15,
  133, 140–41
 power-based analysis and,
  151–52, 153–55
 remedy for violations of, 9–10,
  11–12, 132
 in testimonial phase,
  22–23, 27–28
education and training
 of workers, 52–53,
 107–10, 147
electricity outages, 75–76,
 95–96, 124–25

El Salvador, light manufacturing stakeholder consultation in, 42–43
energy sector, 47–48. *See also* extractive industries
entitlement vehicle phase of stakeholder consultation, 29–35
Equal Exchange, 28–29
extractive industries. *See also* mining industries; *specific companies*
   BHRRC data collection on, 40
   in Bonao, 99–100
   complaints filed in OECD National Contact Point reporting system, 46–47
   corporate disclosure policies in, 47
   damage control phase and, 16–17
   hierarchy of rights in, 152
   mechanisms for stakeholder consultation in, 41–42, 42t, 133–34
   stakeholder consultation in, 3, 12, 42–43, 44
Exxon Valdez oil spill, 19

factories. *See also* Alta Gracia Apparel; Hanes Corporation; light manufacturing industries
   communities affected by, 1, 2–3
   supply chain management and, 31
   WSR and, 13
Fair Food Program (US), 155–56
Fair Labor Association (FLA), 24–25
Falconbridge (mining corporation), 99–100, 112, 114, 126

fatalism, 78, 84–85
Fine, Janice, 154–55
food insecurity, 102
forestry and fishing industries, 47–48
Foucault, Michel, 16
free, prior and informed consent mechanism, 40
Fung, Archon, 150

GATT (General Agreement on Tariffs and Trade), 21
gender issues
   employment opportunity gender gap, 131
   gender-based violence, 152
   gender inequality, 67
   protests based on, 27
   skills training gender gap, 109–10
genealogy of stakeholder consultation, 12, 15–35, 165
   damage control phase, 16–21
   entitlement vehicle phase, 29–35
   testimonial phase, 21–29
General Agreement on Tariffs and Trade (GATT), 21
Georgetown University, 52
global landscape of stakeholder consultation, 12, 36–60
   comparable studies, analysis of, 46–51
   data analysis, 39–46, 42t
   data collection, 56–57
   mapping trends, 37–39
   study methodology, 39–46, 51–55
Goodhart, Michael, 140–41
Gordon, Jennifer, 136, 154–55
governance
   community well-being and, 77–80, 96

corruption and, 74–75, 78–79, 120–21, 127, 131–32, 144
remedy role of government, 145–49
wages and working conditions affected by, 4
grassroots. *See also* Bonao; local contexts of stakeholder consultation; Villa Altagracia
BHR and, 2, 34
case study design and, 13
damage control phase and, 19, 20–21
entitlement vehicle phase and, 34
knowledge of and receptivity to stakeholder dialogue, 83–86, 121–25
NGOs and, 43–44
perceptions of constraints on stakeholder dialogue, 81–83, 118–21
perceptions of responsibility for community well-being, 77–80, 113–18
remedy scope limitations and, 49–50, 86–88, 125–29, 147
Green March Movement (*Movimiento Marcha Verde*), 120–21, 131–32
Guatemala, extractive industries stakeholder consultation in, 42–43

Hanes Corporation plant in Bonao
business model of, 52–53, 98–99
labor rights and, 105–7, 109–10
philanthropic efforts by, 106–7, 124, 130
wages paid by, 103, 129, 147
worker training programs for, 108–9, 129

herding effects, 37
Hertel, Shareen, 27
hierarchy of rights, 151–52
Hopgood, Stephen, 139–40
Human Development Index (HDI), 17–18, 65–66
Human Development Report (UNDP), 17–18, 65–66
human rights. *See also* business and human rights (BHR) framework; economic rights; labor rights
in agriculture industries, 154–56
in entitlement vehicle phase, 31
in extractive industries, 44–46, 152
governance and, 6–7
hierarchy of, 151–52
in light manufacturing industries, 4–5, 53–54, 105–6
policymaking for, 15, 133, 140–41
power-based analysis and, 151–52, 153–55
remedy for violations of, 9–10, 11–12, 132
supply chain management and, 24–25, 31, 44
in testimonial phase of stakeholder dialogue, 22–23, 25, 27–28
in textile industries, 24–25, 53–54, 62
violations approach to monitoring, 10
WSR model and, 139–40

ICERD (International Convention on the Elimination of All Forms of Racial Discrimination), 20–21

ILO (International Labour Organization), 23, 44, 160
IMF (International Monetary Fund), 16–17, 18
impact assessment mechanism, 40
India
    extractive industries stakeholder consultation in, 42–43
    light manufacturing stakeholder consultation in, 42–43
    Bhopal disaster (1984), 19–20
Indonesia
    extractive industries stakeholder consultation in, 42–43
    light manufacturing stakeholder consultation in, 42–43
infrastructure
    in Bonao, 75–76, 103–4, 124–25
    electricity outages, 75–76, 95–96, 124–25
    in Villa Altagracia, 75–77, 95–96
    waste management and disposal, 76–77
    water supply, 40, 41–42, 42t, 76
Instituto Nacional de Formación Técnico Profesional (INFOTEP), 52–53, 57, 108–10, 120, 131, 147
InterAmerican Commission for Human Rights, 20–21
International Convention on the Elimination of All Forms of Racial Discrimination (ICERD), 20–21
International Labor Rights Forum, 141–42
International Labour Organization (ILO), 23, 44, 160
International Monetary Fund (IMF), 16–17, 18
The Invisible Adjustment (UNICEF), 17
Ivory Coast, light manufacturing stakeholder consultation in, 42–43

JavaScript Object Notation (JSON), 39, 40
Jenkin, Eleanor, 152–53
Jolly, Richard, 17–18
juntas de vecinos, 115–19, 122–23, 128–29

Kenya, extractive industries stakeholder consultation in, 42–43
Kingsbury, Benedict, 17–18
Kline, John M., 62, 63–64, 72–73, 82, 129
Knights Apparel, 98–99
knowledge of stakeholder dialogue
    in Bonao, 121–25
    in Villa Altagracia, 83–86
Knuckey, Sarah, 152–53

labor market, 64, 107–10, 132
labor rights
    in agricultural industries, 154–56
    in entitlement vehicle phase, 31
    in extractive industries, 44–46
    in light manufacturing industries, 105–6
    supply chain management and, 24–25, 31, 44

in testimonial phase of
  stakeholder dialogue,
  22–23, 25
in textile industries, 24–25,
  53–54, 62
WSR model and, 139–40
labor unions, 67, 93, 128
Lange, Dorothea, 22–23
Las Delicias neighborhood,
  Bonao, 60, 98–132
  challenges for residents
    in, 102–5
  community attitudes on nature
    of remedy in, 125–29, 147
  constraints on stakeholder
    dialogue in, 118–21
  corruption in, 120–21, 127
  crime in, 104–5
  data collection in, 56–57
  demographics of, 52–53
  electricity outages in, 75–76,
    103–4, 124–25
  food insecurity in, 102
  industrial economy in,
    105–7
  infrastructure challenges
    in, 103–4
  labor market in, 107–10
  lessons learned, 129–31
  local context for, 57–60
  local knowledge of and
    receptivity to stakeholder
    dialogue in, 121–25
  policy implications, 131–32
  power-based analysis in, 154
  responsibility for community
    well-being in, 113–18
  stakeholder perceptions
    in, 98–102
  subjective socioeconomic status
    in, 110–13, 129
  worker training in, 107–10

light manufacturing industries
  BHRRC data collection on, 40
  complaints filed in OECD
    National Contact Point
    reporting system, 46–47
  corporate disclosure policies
    in, 47
  mechanisms for stakeholder
    consultation in, 41–42,
    42t, 133–34
  power-based analysis
    in, 155–56
  stakeholder consultation trends
    in, 42–43, 44
  supply chain and,
    31–32, 59–60
listservs, 118–20
living wage, 52, 53, 59–60,
  62, 63–64, 72–73,
  94–95, 143–44
local contexts of stakeholder
  consultation, 12, 36–60
  comparable studies, analysis
    of, 46–51
  data analysis, 39–46, 42t
  data collection, 56–57
  mapping trends, 37–39
  study methodology,
    39–46, 51–55
Locke, Richard, 31, 45–46, 59–60

managerial capture, 138–39
manufacturing. *See* light
  manufacturing industries
Marchena Hospital (Bonao),
  106–7, 124–25
mechanisms for stakeholder
  consultation
  complaints mechanism, 40,
    41–42, 48–49
  in extractive industries, 41–42,
    42t, 133–34

mechanisms for stakeholder
consultation (*cont.*)
free, prior, and informed
consent mechanism, 40
impact assessment
mechanism, 40
in light manufacturing
industries, 41–42,
42t, 133–34
Melish, Tara, 138–39, 144
MEPyD (*Ministerio de Economía,
Planificación y Desarrollo*),
57, 117
Merry, Sally Engle, 17–18
methodology of stakeholder
consultation study,
39–46, 51–55
questionnaire (English), 169
questionnaire (Spanish), 174
snowball sampling, 56, 100–1
Mexico, light manufacturing
stakeholder consultation
in, 42–43
Milk With Dignity certification
program (US), 155–56
minimum wage, 52, 54–55, 62,
103, 147
mining industries, 47–48,
99–100. *See also* extractive
industries
*Ministerio de Economía,
Planificación y Desarrollo*
(MEPyD), 57, 117
Ministerio de la Mujer (Ministry of
Women), 57
MIT Sloan School and
Management
Review, 33–34
Moleres, Fernando, 23
Morris, Len and
Georgia, 23

Movimiento Marcha Verde
(Green March Movement),
120–21, 131–32
Moyn, Samuel, 139–40
MSI Integrity, 8–9, 47–48, 49–50
Mugler, Johanna, 17–18
multiplier effects, 52, 72–73
multi-stakeholder
initiatives (MSIs)
challenges of, 136–44
drivers for, 8–9
emergence of, 2
government's role in remedy
and, 145–49
industry trends in, 47–48
policy reforms and, 135
power-based analysis
and, 149–57
Murrow, Edward R., 22–23
Mwangi, Wagaki, 7
Myanmar, extractive industries
stakeholder consultation
in, 42–43

NAFTA (North American Free
Trade Agreement), 21
Narayan, Deepa, 30–31
National Consumers
League, 23–24
National Labor
Committee, 28–29
neighborhood associations,
92, 115
Nestlé, 19
Nigeria
extractive industries
stakeholder consultation
in, 42–43
light manufacturing
stakeholder consultation
in, 42–43

nongovernmental organizations
(NGOs). *See also specific
organizations*
community well-being and,
79–80, 115, 117
genealogy of stakeholder
consultation and, 15
human rights advocacy by, 23
MSIs and, 138–39
testimonial phase and, 23–24,
25–26, 27–28
trends in stakeholder
consultation data, 37
non-state-based nonjudicial
remedies, 10–11
North American Free Trade
Agreement (NAFTA), 21

Occupy Wall Street movement, 30
OECD National Contact Point
reporting system, 46–47, 50
OECD Watch, 46, 48–50

Papua New Guinea, gender-based
violence in, 152
Park, Sung-Joon, 17–18
participatory governance, 34–35
Pattipatti, Krishna, 157–59
Peru, extractive industries
stakeholder consultation
in, 42–43
pharmaceuticals industry, 40,
41–42, 42t
philanthropy, 87, 91–92, 106–7,
130, 163–64
Philippines, extractive industries
stakeholder consultation
in, 42–43
policy reform and
recommendations,
13–14, 133–62

in Bonao, 131–32
government's role in
remedy, 145–49
grounded theory on complex
problems, 157–62
MSI vs. WSR
approaches, 136–44
power-based analysis
and, 149–57
Porgera Joint Venture Gold
Mine (Papua New
Guinea), 152–53
poverty
in Dominican Republic, 65–66
in light manufacturing
communities, 4, 133–34
remedy scope and, 13,
146–47, 153
structural roots of, 57–58, 133
subsidies and, 74–75
Poverty Reduction Strategy
Papers (PRSPs), 18, 30–31
power-based analysis, 149–57
power grid outages, 75–76,
95–96, 124–25
problem-driven theory
building, 157–58
public assistance, 72
punitive compliance model, 45–46

receptivity to stakeholder
dialogue
in Bonao, 121–25
in Villa Altagracia, 83–86
redistributive justice, 34
regulatory framework
community well-being and,
77–80, 96
labor laws, 67
remedy role of
government, 145–49

regulatory framework (cont.)
  wages and working conditions affected by, 4
Reinecke, Juliane, 143
remedy
  in Bonao, 125–29, 147
  claim-making and, 9–10
  community attitudes on nature of, 86–88, 125–29
  complaints mechanism and, 48–49
  defined, 9–10
  for economic rights violations, 9–10, 11–12, 132
  entitlement vehicle phase and, 34
  government's role in, 145–49
  poverty and, 13, 146–47, 153
  redistributive justice and, 34
  regulatory framework and, 145–49
  Ruggie Principles and, 10–11
  in Villa Altagracia, 86–88, 146–47
remittances, 72
Rieth, Lothar, 7
Rio Tinto, 150–51
Risse, Thomas, 7–8
Rodríguez-Garavito, César, 138, 149–50
Romano, U. Roberto "Robin," 23
Ropp, Stephen, 7–8
Rottenburg, Richard, 17–18
Rovenpor, Janet, 138–39
Ruggie, John, 29–30, 38
Ruggie Principles, 10–11, 21, 29–30, 136, 149, 150–51

Salgado, Sebastião, 23
Saro-Wiwa, Ken, 19
Satyarthi, Kailash, 160
Schmitz, Hans Peter, 7

Schrank, Andrew, 67, 107–8, 109
Sellers, Sean, 155–56
Sen, Amartya, 34–35, 165
Sethi, Prakash, 138–39
shame-and-blame tactics, 20–21, 26–27
Shell Oil, 19
SHIFT (research group), 47, 49–50, 150–52
Sikkink, Kathryn, 7–8
snowball sampling, 56, 100–1
Snyder, Jack, 17–18
social media, 81–82
Soundararajan, Vivek, 31–33, 34–35
South Africa
  extractive industries stakeholder consultation in, 42–43
  light manufacturing stakeholder consultation in, 42–43
stakeholder consultation
  in Bonao, 98–132 (see also Bonao)
  conceptual framework, 5–12
  constraints on, 81–83, 118–21
  as framework for negotiating risk, 11–12
  genealogy of, 12, 15–35, 165 (See also genealogy of stakeholder consultation)
  global landscape of, 12, 36–60 (See also global landscape of stakeholder consultation)
  knowledge of, 83–86, 121–25
  local contexts of, 12, 36–60 (See also local contexts of stakeholder consultation)
  mechanisms for, 40, 41–42, 42t
  receptivity to, 83–86, 121–25

study methodology, 169, 174, 39–46, 51–55
  in Villa Altagracia, 61–97
  (*See also* Villa Altagracia)
state-based remedies, 10–11. *See also* governance
structural adjustment policies (SAPs), 16–17, 18
subjective social status (SSS)
  in Bonao, 110–13, 129
  case study methodology and, 54–55
  in Villa Altagracia, 64, 80, 96–97
subsidies, 74–75
sunk costs, 3, 44
supply chain management
  global nature of, 1
  in light manufacturing industries, 31–32, 59–60
  power relations and, 150–51
  WSR model and, 136–37
sweatshops, 27–28, 63
systems engineering approach, 157–62

Tanzania, extractive industries stakeholder consultation in, 42–43
teen pregnancy rates, 103–4
Terry, Sara, 23
testimonial phase of stakeholder consultation, 21–29
textile industries. *See also* Alta Gracia Apparel; Hanes Corporation
  labor rights in, 24–25, 53–54, 62
  stakeholder consultation in, 3–4
  training and education of workers, 52–53, 107–10, 147
Transparency International, 131–32

Uganda, extractive industries stakeholder consultation in, 42–43
underemployment, 64, 132
UN Development Programme
  Human Development Report, 17–18, 65–66
unemployment, 64, 132
UN Global Compact, 19–20, 29, 30, 33
UN Guiding Principles on Business and Human Rights (Ruggie Principles), 6–7
unions, 67, 93, 128
United Nations
  business and human rights (BHR) framework and, 1–2, 6–7, 30, 37–39
  Convention Against Corruption, 144
  Convention on Access to Information, Public Participation in Decision-Making and Access to Justice in Environmental Matters, 144
  Convention on the Rights of Persons with Disabilities, 144
  International Convention on the Elimination of All Forms of Racial Discrimination (ICERD), 20–21
  Special Rapporteur on Extreme Poverty and Human Rights, 150
  Special Representative on Business and Human Rights, 29
  testimonial phase and, 25–26
United Nations Children's Fund (UNICEF), 17, 160

United Nations Conference on
  Trade and Development
  (UNCTAD), 136
United Nations Economic
  Commission for Europe, 144
United States
  agricultural labor rights
    in, 154–56
  industrial harm to minority
    communities in, 20–21
United Students Against
  Sweatshops, 63
University of Connecticut
  Business and Human Rights
  Initiative, 13–14, 135–36,
  142, 148

venue-shopping, 20–21
victim-narrative advocacy, 21–29
Villa Altagracia (Dominican
  Republic), 61–97
  challenges for residents in,
    72–77, 90
  community attitudes on nature
    of remedy in, 86–88, 146–47
  commuting costs in,
    68–69, 70–71
  constraints on stakeholder
    dialogue in, 81–83
  corruption in, 74–75
  data collection in, 56–57
  demographics of, 52
  electricity outages in,
    75–76, 95–96
  industrial economy in,
    65–71, 93–94
  infrastructure challenges in,
    75–77, 95–96
  knowledge of stakeholder
    dialogue in, 83–86
  lessons learned, 88–94
  local context for, 57–60

neighborhood associations
  in, 92
policy implications, 94–97
power-based analysis in, 154
receptivity to stakeholder
  dialogue in, 83–86
responsibility for community
  well-being in, 77–80, 91
violence in, 73–74, 154
waste management and
  disposal in, 76–77
water supply in, 76
violations approach to
  economic and social rights
  monitoring, 10
violence
  gender-based, 152
  in Villa Altagracia,
    73–74, 154

Wake Forest Baptist Medical
  Center, 106–7
waste management and
  disposal, 76–77
water supply, 40, 41–42, 42t, 76
Wicks, Andrew C., 32–33, 34–35
WITNESS (advocacy
  organization), 23–24
Wolf, Klaus Dieter, 7
Worker-Driven Social
  Responsibility
  Network, 155–56
worker-driven social
  responsibility (WSR)
  paradigm
  challenges of, 136–44
  government's role in remedy
    and, 145–49
  human rights and, 139–40
  policy reforms and, 135
  power-based analysis
    and, 149–57

remedy and, 2
supply chain management
   and, 136–37
in Villa Altagracia, 52, 59–60, 88,
   95, 146–47, 155–56, 163–64
worker education and training,
   52–53, 107–10, 147.
   See also *Instituto Nacional de
   Formación Técnico Profesional*
   (INFOTEP)

Workers' Rights Consortium
   (WRC), 24–25, 62, 63,
   105–6, 109–10,
   136–37, 141–42
World Bank, 16–17, 18, 117
World Trade Organization (WTO),
   21, 27–28
Wright, Erik Olin, 150